CONTENTS

CHRONOLOGICAL LIST OF HEARINGS

2004

THURSDAY, JUNE 24, 2004

CONTRACTOR SUPPORT IN THE DEPARTMENT OF DEFENSE

STATEMENTS PRESENTED BY MEMBERS OF CONGRESS

WITNESSES

APPENDIX

[H.A.S.C. No. 108–34]

CONTRACTOR SUPPORT IN THE DEPARTMENT OF DEFENSE

HEARING

BEFORE THE

READINESS SUBCOMMITTEE

OF THE

COMMITTEE ON ARMED SERVICES
HOUSE OF REPRESENTATIVES

ONE HUNDRED EIGHTH CONGRESS

SECOND SESSION

HEARING HELD
JUNE 24, 2004

U.S. GOVERNMENT PRINTING OFFICE

98–853 WASHINGTON : 2005

For sale by the Superintendent of Documents, U.S. Government Printing Office
Internet: bookstore.gpo.gov Phone: toll free (866) 512–1800; DC area (202) 512–1800
Fax: (202) 512–2250 Mail: Stop SSOP, Washington, DC 20402–0001

READINESS SUBCOMMITTEE

JOEL HEFLEY, Colorado, *Chairman*

HOWARD P. "BUCK" McKEON, California
JOHN N. HOSTETTLER, Indiana
WALTER B. JONES, North Carolina
JIM RYUN, Kansas
ROBIN HAYES, North Carolina
HEATHER WILSON, New Mexico
KEN CALVERT, California
J. RANDY FORBES, Virginia
JEFF MILLER, Florida
TOM COLE, Oklahoma
ROB BISHOP, Utah
CANDICE S. MILLER, Michigan
MIKE ROGERS, Alabama
TRENT FRANKS, Arizona
JOHN M. McHUGH, New York

SOLOMON P. ORTIZ, Texas
LANE EVANS, Illinois
GENE TAYLOR, Mississippi
NEIL ABERCROMBIE, Hawaii
SILVESTRE REYES, Texas
VIC SNYDER, Arkansas
CIRO D. RODRIGUEZ, Texas
ROBERT A. BRADY, Pennsylvania
BARON P. HILL, Indiana
JOHN B. LARSON, Connecticut
SUSAN A. DAVIS, California
RICK LARSEN, Washington
JIM MARSHALL, Georgia
MADELEINE Z. BORDALLO, Guam

MARY ELLEN FRASER, *Counsel*
DIANE BOWMAN, *Staff Assistant*

CONTRACTOR SUPPORT IN THE DEPARTMENT OF DEFENSE

HOUSE OF REPRESENTATIVES,
COMMITTEE ON ARMED SERVICES,
READINESS SUBCOMMITTEE,
Washington, DC, Thursday, June 24, 2004.

The subcommittee met, pursuant to call, at 10:01 a.m., in room 2118, Rayburn House Office Building, Hon. Joel Hefley (chairman of the subcommittee) presiding.

OPENING STATEMENT OF HON. JOEL HEFLEY, A REPRESENTATIVE FROM COLORADO, CHAIRMAN, READINESS SUBCOMMITTEE

Mr. HEFLEY. Good morning, and welcome to this morning's Readiness Subcommittee hearing on the topic of contractor support in the Department of Defense.

The Department of Defense relies on three major workforce groups to perform its mission. First and primary is the military personnel, including active duty, reservists, and the guard. The second is the civilian workforce; and the third, of course, the focus of this hearing, is the contractor workforce. This latter group provides a significant level of effort to the Department of Defense.

Support or service contractors provide a myriad of services, including maintenance of weapon systems, logistics support, base operating support, linguists and operation of communication networks, to name just a few. The Department of Defense policy is to rely on the most effective mix of active, reserve, civilian and contract resources to fulfill peacetime and wartime missions.

In fiscal year 2003, the Department procured approximately $209 billion of equipment, items and services. Of this amount, $90.5 billion was for supplies and equipment, $76.2 billion for services, $33.1 billion for research and development, and $9.2 billion for construction. The focus of the hearing will be on the range of services purchased with the $76.2 billion and the management and oversight of these services.

Many of the questions the Members have today came to light as the use of contractors in Iraq became apparent. The subcommittee is interested in how the Department procures such services, the policy on management and oversight, and whether the Department should change some of these policies. Finally, what are the lessons learned from Iraq with respect to the use of contractors on the battlefield?

I look forward to hearing from each of the witnesses, and now I am going to turn to Solomon Ortiz, who is my partner here, for any opening statement he might want to make.

Mr. Ortiz.

STATEMENT OF HON. SOLOMON P. ORTIZ, A REPRESENTATIVE FROM TEXAS, RANKING MEMBER, READINESS SUBCOMMITTEE

Mr. ORTIZ. Thank you very much, Mr. Chairman. I join you in welcoming our distinguished panel of witnesses, and I thank you all for being here with us this morning.

The nature of this hearing is very timely. Contractors provide valuable services to the Department of Defense. They are one leg of a tripod of manpower that includes military personnel and government employees. They maintain equipment and weapons systems, they supply our forces and deliver critical logistical capabilities, and they provide for the well being and quality of life of our servicemembers and their families at home and abroad.

In some cases, unique skills critical to mission success reside only in the ranks of our contractors. In short, without them, large portions of our military would not function and important missions would go unaccomplished. These contractors are fine, committed Americans in every sense; and we are very grateful for their efforts.

Still, I have some concerns about the roles contractors play in providing essential goods and services for our fighting men and women. Our readiness depends upon our ability to meet the demands of our forces during peak times and to preserve our industrial base to respond rapidly to demands during times of relative quiet.

Contractors, by their very nature as private enterprises, are subject to the whims of the marketplace. If we unexpectedly require a key capability that has lain dormant for some time, we must be positive that it will be there when needed. There is no room for error, and I want to be clear that cost is not the only or best metric for success.

To that end then, I want to be sure that I understand how decisions such as how and when contractors are used in lieu of military or government employees are made and how they are supervised.

Furthermore, there seems to be a push to contract out services that have traditionally resided in government-owned entities. The process of converting government functions to contractors is difficult. There is a potential for serious problems in the process and I would advocate a go-slow approach whenever we consider it.

Additionally, watching the situation in Iraq develop, it is clear that the Department does not have its act completely together with regard to managing contractors. Clearly, the nature of trying to conduct business in a very difficult wartime setting is challenging, and I would not minimize the difficulties facing our people in Iraq. Still, we have cost overruns and services interrupted.

We see clear indications that oversight is lacking; and a comprehensive plan for awarding, managing and monitoring contracts is not sufficiently in place. If this is a case study of the way the Department is organized to integrate contractors into the overall structure of the Department, then I see some reasons for concern.

Members of the panel, I looked over the testimony and responses to our questions. Nevertheless, Mr. Chairman, I think these are issues surrounding contracting, and particularly with regards to activities in Iraq, we will not have sufficient time here today. This is why I recommend we also address these issues in a full committee hearing sometime in the near future, Mr. Chairman. And thank you for the time.

Mr. HEFLEY. Thank you, Mr. Ortiz.

We have with us today the Honorable Michael Wynne, who is Under Secretary of Acquisition Technology and Logistics. I guess you are acting Under Secretary still, Mr. Wynne?

Secretary WYNNE. I send in my dues every month, Mr. Chairman.

Mr. HEFLEY. That is good. That is good. Well, we are delighted to have you here. And you are accompanied today by Lieutenant General C.V. Christianson, who is the Deputy Chief of Staff for G4, Department of the Army; and I don't know if you have a statement or not or if you are just here to answer questions, but whichever is fine with us.

Also the Honorable John Young Jr., the Assistant Secretary of the Navy for Research, Development and Acquisition, from the Department of the Navy; the Honorable Marvin Sambur, Assistant Secretary of the Air Force for Acquisition; and Ms. Tina Ballard, Deputy Assistant Secretary of the Army for Policy and Procurement. And we are delighted to have each of you here.

And let me say at the outset that your complete statements, without objection, will be made a part of the record, so if you would summarize your statements in whichever way you would like to, we will turn it over to you, Mr. Wynne.

STATEMENT OF HON. MICHAEL W. WYNNE, ACTING UNDER SECRETARY OF DEFENSE, ACQUISITION, TECHNOLOGY & LOGISTICS DEPARTMENT OF DEFENSE

Secretary WYNNE. Well, thank you very much, Mr. Chairman, Mr. Ortiz; it is good to be here. Members of the committee, I appreciate your coming. Thank you for the opportunity to discuss the role of contractors within the context of the Department of Defense.

America's defense mission is clear. Our troops must protect America and her interests wherever and whenever they are called. This requires that our military and defense organizations use civilian contractors to allow the war fighter to concentrate on his or her core mission.

This relationship is not a new one. We have been relying on defense contractors since General George Washington used them as wagon drivers to haul supplies. Today, in the Middle East, contractors provide maintenance for high-tech equipment and construction services, food, water and more. Back home the contractors play a direct role in the research and development of major weapons systems, provide critical support for our military bases, from high-tech nano-engineering on classified programs, to hauling trash and providing telephone service.

Our budget for contracting goods and services over the past five years has moved more and more toward service contracts and away from simply hard goods, as we try to combine the best elements of commercial practice. This change is a reflection of our desire to buy capabilities. When we buy capabilities, contracts that once fell under the heading of goods are now considered service contracts.

For example, in information technology, in the past, contracts were for the purchase of a certain number of computer terminals and a certain number of software licenses, which are considered both, in their way, goods. And we would have had to put out a new contract for computer hardware with each upgrade that came along and perhaps order up a new license. This was a very slow and expensive process that left us behind the information technology curve.

Today, we put out a contract for access to and use of the latest and most up-to-date information technology; so it is now up to the contractor to both purchase and supply both the hardware and software, complete with licenses, that we need. This makes the former contract for goods, in fact, turn into a service contract.

This new direction in contracting requires us to rewrite our policies to mirror this new reality. So we have developed a management structure to manage service contracts, and we put that out about two or three years ago under the aegis of Secretary Pete Aldridge in managing service contracts that came in over a certain dollar threshold.

Next, we are developing a DOD directorate for management of contractor personnel during contingency operations to address the issues of contractor personnel management. Under this policy, we will create a single DOD directive, as recommended by the General Accounting Office last June; and we will capture the lessons learned from recent operations, frankly, building on the Army's practices that have been in place for some time.

Second, we are working on a rule to ensure the uniform treatment of contractors who accompany a deployed force and to enable the combatant commanders to rapidly adjust to changing conditions on the battlefield through cross-communications that are required.

In addition to these new rules, we expect to issue new guidance soon to the DOD contracting officers for oversight of wartime contracts. Under this new governance, DOD contracting officers would have to approve the use of non-DOD contract vehicles. We think that these policies and procedures that we have learned over time are going to really help us.

As we move forward in this ever-evolving process of contracting, let us not lose sight of the true heroism and patriotism that Mr. Ortiz had mentioned for these civilian contractors, 30 plus of whom have died in the Middle East in support of our armed forces over the past year. Their persistence, their bravery and their courage under fire should be recognized, and they are still with us, Mr. Chairman, something we are very proud of.

We must look for ways to make this system work better so that the DOD, the contractors, and frankly all of America continue to benefit from this methodology. I very much appreciate the support that the House Armed Services Committee has given DOD. I applaud you on your recent passage of your authorization bill for this year, as well as to the war fighters that you always are quick to recommend and quick to recognize.

I look forward to continuing to work together. I will turn it over now to my colleague, John Young, for his opening statement.

[The prepared statement of Secretary Wynne can be found in the Appendix on page 31.]

Mr. HEFLEY. Mr. Young.

STATEMENT OF HON. JOHN J. YOUNG, JR., ASSISTANT SECRETARY OF THE NAVY (RESEARCH, DEVELOPMENT & ACQUISITION) DEPARTMENT OF THE NAVY

Secretary YOUNG. Mr. Chairman, distinguished members of the subcommittee, good morning. Thank you for this opportunity to discuss the Department of the Navy's policies on contractor support for the war fighter and our practices in overseeing those contracts.

The Navy and Marine Corps Team is unique among the services in that our mission, being expeditionary in nature, is focused on operating independently; and our operating culture is based on being able to sustain operations organically in an isolated environment with limited foreign contractor support. We are often asked to operate in areas where access may be restricted.

Naval forces must be capable of initiating and sustaining combat operations by sea, land and air without being limited by a lack of logistics or host nation support. The Navy does not carry a large cadre of contractors aboard ship, and the Marines do not typically take contractors ashore to support their landing forces.

Throughout the history of the naval services, we have relied to some degree on the support of contractors to effectively carry out our mission; but, historically, the Navy and Marine Corps have relied on organic capabilities to sustain operations. We acquired, stocked and distributed an inventory of spare parts, necessitating maintenance of an extended logistics supply chain. Over the past 10 to 15 years, we have been moving away from our own logistics support agent concept, using an appropriate balance of government and industry to deliver faster results and greater operational availability.

Presently, our industry partners are supporting our forward-deployed military forces primarily in two areas: weapons systems support and emergency construction. Our high-tech weapons systems frequently need the expertise only contractors can provide in order to keep them operating at the desired levels of performance. We contract for field technical services when these services aren't already covered by the production contract or a negotiated logistics support contract.

The performance of the work is monitored by either the Defense Contract Management Agency or the Navy or Marine Corps end user at the site where the field technical support is performed. In either case, verification of successful performance of the work is a necessary prerequisite for contractor payment.

For construction, the Naval Facilities Engineering Command, or NAVFAC, uses the competitively awarded Emergency Construction Capabilities, or CONCAP, contract, which provides the Navy an immediate response for construction and related engineering services in support of contingency and emergency situations.

NAVFAC maintains the organic capability to manage and oversee all its construction work, either through its Resident Officer in

Charge of Construction organization or through the assignment of a Navy technical representative in locations where ROICCs aren't available. Contractor payments are handled through this organizational structure, again, with payment authorized only after confirmation that the work has been satisfactorily completed.

Mr. Chairman, in summary, our industry partners provide essential support when necessary to our deployed forces that enables our forces to focus on their core national defense mission. We contract with them in complete accordance with law and regulation as well as good business judgment. They have truly been effective partners in many instances, going into harm's way and providing our forces the support needed to accomplish the mission whenever and wherever needed. Thank you again for the chance to appear today.

[The prepared statement of Secretary Young can be found in the Appendix on page 40.]

STATEMENT OF HON. MARVIN SAMBUR, ASSISTANT SECRETARY OF THE AIR FORCE, (ACQUISITION) DEPARTMENT OF THE AIR FORCE

Secretary SAMBUR. Mr. Chairman, Mr. Ortiz and members of the committee, in order to give you extra time to ask us questions, I would like to just summarize my opening by making three points.

First, our experience with contractors and contracting has been largely positive. Second, the Air Force has already recognized the growing importance of proper management and oversight of service acquisitions; and in 2002, we established what we now call the Program Executive Office for Combat Mission Support to provide oversight and management of service acquisitions of over $100 million. This particular organization has been lauded by DOD and the Inspector General. So it has been very successful.

Third, and most importantly, we are committed to work closely with OSD and Congress to make whatever corrections are necessary.

Thank you.

[The prepared statement of Secretary Sambur can be found in the Appendix on page 45.]

STATEMENT OF TINA BALLARD, DEPUTY ASSISTANT SECRETARY OF THE ARMY, (POLICY AND PROCUREMENT) DEPARTMENT OF THE ARMY

Secretary BALLARD. Mr. Chairman, Congressman Ortiz and distinguished members of the Subcommittee on Readiness, thank you for this opportunity to report to you on the roles and missions of contractors that support the Department of Defense and the military services, specifically the United States Army.

On May 21 of 2003, the Deputy Secretary of Defense designated the Secretary of the Army as the Department of Defense executive agent for the Office of Reconstruction and Humanitarian Assistance, later to become the Coalition Provisional Authority (CPA) in Iraq, with responsibility to provide administrative, logistics and contracting support for the relief and reconstruction for the people of Iraq.

On January 14 of this year, the Deputy Secretary of Defense further assigned responsibility for acquisition and program management support for CPA to the Secretary of the Army to include all aspects of contracting and program management.

Civilian contractors have become an indispensable part of our Nation's war fighting and peacekeeping capability. With this in mind, the Army has taken a number of important steps to manage our contractors in the field to ensure that they are both effective and economical. I will provide a few examples.

The Army is the first military service to provide a contract clause dealing with contractors accompanying the force, providing consistent guidance across all Army contracts. And the Army's continued efforts to improve management of service acquisitions, the Army Services Strategy Panel, are focused on the importance of developing and maintaining sound acquisition strategies and making smart business arrangements.

And also, we have contracting officer representatives who are authorized to represent a contracting officer. They assist in technical monitoring or administration of a contract, and they are government employees.

I have a deep appreciation for the courage, competence and dedication, the hard work of our contracting personnel. Their work has been and continues to be performed in accordance with proper procedures by military servicemembers and civilians who are concerned and committed to accomplishing their work under the toughest and most austere conditions.

Our contractors and contracting personnel are often the unsung heroes of any deployment. Their customers are the war fighters, the men and women who depend on them to do their jobs.

Thank you. I look forward to your questions.

[The prepared statement of Secretary Ballard can be found in the Appendix on page 50.]

Mr. HEFLEY. Just a quick question or two, and then we will go to our Members here. What is the current policy with respect to determining what functions should or should not be performed by a contractor? And what are your criteria for determining whether it will be a no-bid contract when you do decide that this ought to be contracted out? And how extensive is no-bid contracting?

Mr. Wynne or whoever?

Secretary WYNNE. Let me start by saying that we, over the years, have constructed various definitions of core versus non-core performance. We have also tried to parse it out by government versus non-government. Those two are not coequal definitions; and they have led us down a path of trying to define, during peacetime at least, what would constitute a nominally core and yet should remain the aegis of the government.

For example, things like contracting, specifically taking ownership of the intellectual capital of how a program should be exercised while using the brute strength, if you will, of a contractor system to extend the reach of that contracting officer.

As time goes on in wartime, we also have the efficacies of protection, i.e. at some point in time when you contract for truck drivers, as General Washington did and as we did in Iraq, many times we prohibited their flow north of the Kuwaiti-Iraqi border while we felt it was unsafe activity. As the situation changed, we then allowed civilian contractors to, in fact, drive the trucks throughout to deliver fuel, et cetera, north of the border. That has ebbed and flowed, and it is at the commander's discretion as to how he feels the services should be rendered.

This time within the context of Iraq, we have a very interesting situation where we are restoring stable governance. We are working in concert, in contacts with the U.S. Agency for International Development (USAID) as well as some Iraqi companies. We are inviting investment from outsiders, which all tend to congeal and look the same but they are not the same. Those contractors are supporting us, as Secretary Ballard has pointed out, in fact, have clauses within their contracts making it very clear as to how their actions should be within the context of accompanying forces and also in reporting mechanisms.

We do, however, use performance contracting. Performance contracting means that we are asking for a certain electricity flow or a certain oil flow or a certain water delivery or tons delivered; and that means that the contractors are expected and, in fact, asked to provide whatever necessary security to accomplish that part of that mission. And they do. And that, I would tell you, is one of the harder things to consider as we go from force protection of our own forces to force protection as you might see it in contractors.

But we have very clear definitions, and it is based on the commander's intent and the intent of the commander in the field as to what he constitutes or what he sees as his ability to have organic support versus non-organic support.

As far as the question on no-bid, I would say that we attempt to competitively bid everything that we can. However, we have vehicles that are, in fact, in place for many of the expertise that we need; and those contracts were arrived at on a competitive basis. And so we offer to other agencies the ability to either pile on to our contracts, and they also offer us the ability to pile on to their contracts. And in that regard, we try to stay within the lines; and I think we are learning that we might need to advise the other agencies where they have contracts in place for the expertise we need. But we still need to enforce the DOD-specific regulations and rules, and I understand that that definition is being crisped up and the lessons learned are being prepared.

Goods are not often offered, and when they are, it is justified with a justification assessment that is approved by fairly high levels within the acquisition community and the milestone decision authority. It is not taken lightly, and we do reserve the right, if you will, to take a good hard look at things. Thank you, sir.

Mr. HEFLEY. Thank you.

Mr. Ortiz.

Mr. ORTIZ. Thank you, Mr. Chairman. I think in the year 2003, the General Accounting Office recommended the Department improve its management and oversight of contractors; and they came down with some recommendations. I am pretty sure you are familiar with the recommendations they came up with. But one of the things that bothers me, and maybe you are doing it now, is that a lot of the National Guard and Reserves with special skills are being activated; and I hope that we don't have a duplication of trying to hire private contractors who have the same skills and then we are activating Reserves and National Guard.

How do you screen them out so that we won't have any duplication, because I know that we are now heavily dependent on the National Guard and Reserves, especially on the special skills.

Secretary WYNNE. We are, in fact, heavily dependent upon the National Guard and Reserve; and we are very proud of their accomplishments and the skills that they are in, and they train very well. That having been said, we follow the requirements coming in from the field, from the combatant commanders, when we assess whether or not we have a lack of skill sets available to us.

Just as a simple example, stevedores, we have tremendous capability within our ranks on stevedores. However, the capacity is required to become extraordinary; we will use that as a core capability and extend their reach, if you will, by hiring civilian contractors to supplement them. That does not mean that we do not want to have a core capability in a very interesting field of stevedoring and port management within the context of the Navy; and I am sure my colleague, John Young, will ratify some of that.

The expertise, though, as we distribute them, we identify clear needs for capabilities, and we do examine what the capabilities are within our own force structure first. That having been said, we find ourselves lacking. In spite of our best efforts, we are finding ourselves having to convert, for example, artillerymen into military police. We know that is a significant skill that is in the Guard and Reserves; and yet it is overextended. We respect that.

Mr. ORTIZ. So you feel there is no duplication calling people with the same MOS when they——

Secretary WYNNE. I can't tell you, sir, but the opinion of the Department of Defense is that there is not but one or two instances of duplication. It is not intentional, and it is certainly not done with any plan in mind. It is strictly the fact that perhaps they are either going off roster or coming on roster, and the skills were needed at the time they were ordered.

Secretary YOUNG. Mr. Ortiz, if I might, I am not sure if I understand the question completely, but if I could use an example. We are activating our Reserve component transportation units at the same time we are contracting for transportation capability. One of the reasons you will see that duplication is because we don't want to activate more transportation that we have to, and in some cases we have contracted drivers to operate Reserve component trucks that we have left in theater while we have brought the Reserve component units home.

So if the question is do we have contracted the same skills that are activating, the answer is, yes, in many cases we are. If the next question is why are we doing that, primarily we are doing that to relieve the pressure on the Reserve components to have to continue to respond and support this long-term mission.

Mr. ORTIZ. And another thing that I am concerned about now, what kind of training, or once you contract them out, do you think they are adequately trained to go into harm's way? Because the soldiers, whoever they are, their MOS is to be an infantryman and to be able to fight. What about the contractors? Do you feel that they are adequately trained to go into harm's way?

Secretary YOUNG. The initial training provided for the drivers that we used before we attacked Iraq would not have been adequate to operate in Iraq today. That training program has been changed by the contractors, number one. And number two, the philosophy that forward commanders have today is to integrate the contractors with military forces. So when the first contracted trucks were attacked as they went into Iraq about three weeks after the operation started, we changed the concept of how we would move things forward; and we integrated those trucks with our trucks by setting a separate contracted operation north which was the original plan.

So they have modified their procedures in the theater. They have changed the training for the truck drivers themselves, the contracted drivers; but they are not trained as soldiers. There is not enough time to do that. So what our plan is, and our concept is, is to keep them integrated with the military so they work together.

Mr. ORTIZ. Any idea how many contractors have been killed in this war, in Afghanistan and Iraq?

Secretary WYNNE. Near as we can tell, sir, there are in excess of 50. We think that approximately two-thirds of that number have been actually killed in what would be called military style actions, and one-third probably could be registered as traffic or mishaps or non-battle-related casualties. There are an additional many dozens that have been wounded; and, I think, have all represented America very, very well.

Mr. ORTIZ. Thank you, Mr. Chairman.

Mr. HEFLEY. Mr. McHugh.

Mr. MCHUGH. Thank you, Mr. Chairman. Welcome to you all. I would like to pursue the question that the general and secretary were just alluding to with respect to Iraq specifically—and Afghanistan now that I think about it.

Obviously, these are ongoing missions, but one of the things we all try to do when you are engaged in those kinds of events is to learn some lessons and to modify your procedures as you go along. General Christianson mentioned in the Army's case how you are now integrating trucking contractors in with military convoys. That is certainly one lesson learned. But I am curious, have we made any other adjustments, any surprises to the good or not so good that have caused you to operate your contracting procedures a little bit differently because of what you encountered in either of those theaters?

Secretary WYNNE. One thing that I would say is the Army had both a field manual and a series of contract clauses that they applied from day one into contracts, not knowing what the security situation would, in fact, be in either of those theaters. We all postulated, as you know, that they would be getting better on a quicker pace.

So we embarked on reconstruction activities that were more than the skill level that we have in our armed forces would allow us. We are, in fact, in every case, from operations through reconstruction, through the operation of the coalition provisional authority, by the way, putting together lessons learned teams to bring those things back. We also have heard from folks we sent over there who are in the IG and the GAO and are working with those lessons learned teams.

One of the feedbacks from that is maybe we need to take and build upon the Army's situation for and on behalf of the total DOD, be it Navy, Army, Marine Corps, perhaps even the National Guard, and the Air Force. So that, I would say, is something I think is going to come back to be positive.

The other thing we found out is that—and Clausewitz said it best, when the first bullet whizzes by, strategic plans become all tactical. So we are actually fostering a far better approach to capturing contract activities with contingency-based contracting rules and contingency-based contracting to provide big reachback to the United States or to some other command structure to help these people document their activities and make sure that we have traceability throughout the process.

We sent a whole team over there to assist with that process; and the other thing we are doing is we have now a distribution operations center which is theater based, which now tracks goods, if you will, through the system to their receiving unit to avoid the replication of ordering, trying to make sure that we don't overburden the transportation problem that the Army already has. I think I would turn to my colleague from the Army and see if there is anything else.

Secretary BALLARD. In the Army, we have been collecting several lessons learned associated with this mission. We have sent teams in, as Mr. Wynne indicated, to look at how we have executed the contracts on the ground, brought those lessons learned back. Additionally, we have formed a team of folks in the Army who are looking at how we will execute contingency missions in the future, from how we embed contracting personnel with the units to how we execute contingency missions like this one in the future.

Mr. MCHUGH. I appreciate that. It feels as though it is a work in progress, which I guess is fully understandable.

Rather than pursuing that any further, let me just quickly go over to Secretary Young. Mr. Secretary, you mentioned in your written testimony about the husbanding contracts that the Navy maintains at ports that you may or may not visit so that you don't have to establish a continual Navy presence there. I am assuming, but I shouldn't do that, these contracts are generally contracts given to foreign businesses or do you contract out at such an extent that U.S. firms go to wherever you may be going in anticipation of your arrival?

Secretary YOUNG. I can add some additional detail to the record; but, in general, those are longer term contracts competitively awarded, and U.S. firms do vie, because of the duration, to come and provide those services, the husbanding services.

Mr. MCHUGH. So they would be there on a permanent basis?

Secretary YOUNG. Well, they are contracts put in place in typical ports of call. They are not necessary in ports of call that may be the location for contingency operations.

Mr. MCHUGH. Well, the reason I am asking is because you say, "In lieu of establishing organic Navy presence in every port, we might need to visit . . .," but what I am hearing you say is these are generally at places where you do have regular ports of call.

Secretary YOUNG. Yes.

Mr. MCHUGH. Okay. That clears it up. Thank you very much.

Thank you, Mr. Chairman.

Mr. HEFLEY. Mr. Reyes.

Dr. Snyder.

Dr. SNYDER. Thank you, Mr. Chairman.

Mr. Secretary, Secretary Wynne, I wanted to ask you about what kind of problems you have had. I have looked through the opening statements and I think they give an overview of the work that you do, but I don't think they outline very well any potential problems that you are having that perhaps Congress could help with or that the American people would be interested in. Could you just tick off for me please the 5 or 8 or 10 problems with contracts you have had?

Secretary WYNNE. I think they are all pretty much subject to a Defense Contract Audit Agency (DCAA) audit and sometimes Inspector General activity, and they are fairly well publicized. In fact, we have become sort of targets in the press, if you will, for that purpose.

That having been said, none of the audits have been closed, and many times I have looked and I asked for best practice because you kind of have to ask the question, what is the difference between a cost question, for example, or payments questioned between domestic contracts and contracts in contingency zones like Iraq and Afghanistan? Because what I am trying to do is reach for a standard that I can cohere to.

Dr. SNYDER. Mr. Secretary, I only have limited time. Could you just give me some examples of areas where you have had problems with contracting?

Secretary WYNNE. I don't know that I have had problems. The Army is reporting absolute appreciation for the Logistics Civilian Augmentation Program (LOGCAP) contract that has recently been questioned by the DCAA. The oil is in fact flowing extraordinarily well on behalf of all the folks at Haliburton.

Dr. SNYDER. Well, let me pose it as a hypothetical question. Let's suppose that you had a bad actor, a bad contractor. What is your recourse? Is it the False Claims Act?

Secretary WYNNE. We have first recourse through their contract. We also have recourse through criminal activity if it has to do with fraud. They are subject to U.S. law, and the False Claims Act applies across the board.

Dr. SNYDER. Over the last year and a half or since the beginning of the Iraq war, have we had any criminal charges or False Claim Act charges made against the contractor?

Secretary WYNNE. We have had allegations, none proven thus far. They are all subject to criminal review and the IG has not reported. And as I mentioned, the——

Dr. SNYDER. So no criminal charges and no false claims.

Secretary WYNNE. There have also been false claims alleged through the Defense Contract Audit Agency, but none of those audits have been concluded conclusively; and it is within our context and our system that everybody gets a fair hearing.

Dr. SNYDER. And I agree with that.

Secretary Ballard, in your statement, if I am reading the numbers right, the Army's had 2,800 contracts supporting Iraq, totaling $11.7 billion and 65 personnel. Now, tell me what those 65 personnel do? I mean it seems like 65 personnel, if I divide that up, that is one person, I would assume, managing a tremendous amount of money. And is 65 personnel monitoring $11.7 billion in support of Iraq sufficient, given, I would think, just the logistics of figuring out how to keep a phone line open? Is that sufficient to manage that kind of volume of money in what I think would be a fairly free wielding environment, in terms of watching out for the $11.7 billion taxpayer dollars?

Secretary BALLARD. Sir, part of our strategy has been to keep the footprint forward in Iraq small because of the volatile environment and the security circumstances there. So 65 people forward are actually leveraging the entire acquisition workforce in the rear, not just of the Army but also of the Navy and the Air Force, whom we rely upon to support us in these contract actions as well. So that is how we are able to execute that particular mission.

Dr. SNYDER. Well, I am not sure that that answers the question of a concerned taxpayer who wants to know who is watching out for the $11.7 billion from your statement that are in Iraq. I mean, the Little Rock Police Force is certainly substantially bigger than 65 personnel; and that is a much smaller geographic area. Maybe I am not understanding what these 65 personnel are doing or supposed to do.

Secretary BALLARD. In addition to the 65 personnel that are in our contract office in Iraq, we have Defense Contract Management Agency (DCMA) Iraq, we have DCA auditors who are also forward, the CPA IG, the DOD IG, and the GAO. There are several activities who are forward who are helping us both oversee these contracts as well as administer and manage these contracts, in addition to leveraging the rear to oversee these contracts.

Dr. SNYDER. Mr. Chairman, may I ask one more question?

If you see a press report that—Secretary Wynne, let me address you. If you see a press report, and they come out every once in a while, or you get an email from one of the troops, which happens to us, that says, "We saw this," or let's suppose you have a press report that makes allegations of fraud or waste or a problem. How do you respond to that?

Secretary WYNNE. The first thing that I do, sir, is contact the Inspector General to make sure that we can get the claim. It doesn't rest very long on my desk. If there is an allegation of fraud, I think it is incumbent upon me to further investigate it and not let it rest.

I will say that in the one instance of the LOGCAP contract, I think the extent of the application of that contract got beyond the Kellog, Brown & Root (KBR) and Haliburton management structure; and I know they have beefed that up quite a bit over the course of the time they have been there. I mean they are learning the lessons as well as we have. But any allegation that comes to my desk, sir, I immediately refer.

The other thing that happens is that on a major cost question, we do withhold payment to protect the American taxpayer. So we take action that may run a little counter, if you will, to everybody gets a fair hearing, but, in fact, brings up short because we do defend our auditors as well.

Mr. MCHUGH. Thank you, Mr. Chairman, for your indulgence.

Mr. HEFLEY. Mr. Reyes.

Mr. REYES. Thank you, Mr. Chairman. I apologize for having to leave. I have another meeting going on at the same time. And I don't know if this was asked but the question I have, essentially, CPA Order 17 is a status of forces agreement, and provides immunity from Iraqi laws for U.S. forces and officials as well as for contractors and their employees. Unfortunately, however, it remains unclear to Special Agent in Charge (SAIC) as to whether that order will be legally defensible when on the 30th of June the CPA is dissolved and sovereignty is turned over to the interim Iraqi government.

Two questions: What is the status of forces agreement with respect to Iraq? And, second, if it is not finalized by midweek next week when the actual transfer of power occurs, what kind of a position will we leave U.S. contractors to face the potential risks in Iraq?

Secretary WYNNE. Can I start on that, sir, and maybe——

Mr. REYES. Sure, absolutely.

Secretary WYNNE [continuing]. And maybe turn to my colleague, Secretary Ballard? The status of forces agreement that was in place through CPA number 17 has been negotiated with the interim government so that it will be in effect, if you will, on 12:01, on the 30th of June, or the 1st of July. Following that time, we are then subject to the same, trade craft and state craft that our embassy has managed in Afghanistan and other places where our forces are, which normally I think is a maintenance of agreed to activities.

So we don't forecast there will be tremendous change. That does not say that there won't be some pressure on the interim government to do something, but I think right now all the indications are they are going to live up to their agreements.

Secretary BALLARD. On March 8, Iraqi governing council established transitional administrative law in Iraq; and in that law it states that the orders written and established by Ambassador Bremer will continue to have full force in effect after June 30. So it is our interpretation then that CPA Order 17 continues to apply after June 30 in accordance with that transitional administrative law.

Mr. REYES. If there are no more comments, one issue that I have not received a response from, and this is predicated from concerns that have been expressed to me by sometimes parents and sometimes spouses of contractors that are actually working in Iraq for Brown and Root or Haliburton largely in those convoys. Are they allowed now to carry weapons, because based on the questions that I have been asked, the general consensus is that contractor employees in those situations are prohibited from being armed because of, I guess, violation of Iraqi law?

Secretary WYNNE. Actually, I believe that it is up to the component commander as to whether he sees the necessity for being armed. However, you should be aware that many of the contracts call for security of their personnel; and the contractor, therefore, can turn to a security-providing group to assist them. But to my knowledge, there are some times that we have restricted contractors from carrying weapons dominantly when we have our soldiers right next to them and we don't want any undue fratricide.

Let me turn to Secretary Ballard.

Secretary BALLARD. The combatant commanders make the determination as to whether or not contractors can carry weapons. It is my understanding that, based on the information you have here at this time, the combatant commanders are reviewing that policy specifically in relation to convoys.

Mr. REYES. They are reviewing that policy?

Secretary BALLARD. Yes, sir.

Mr. REYES. Well, you know, that policy has been under review now—because we had General Meyer here well over a month ago, I guess. And that was the response then. How long do you anticipate the review process to take place?

And I say that with all due respect because mothers, fathers, and children continue to ask what our policy is, because those convoys are one of the most dangerous types of jobs that those contractors are doing; and, again, they are not allowed to carry weapons, presumably because there is an armed guard or armed vehicles escorting that convoy. However, when they get attacked and it starts hitting the fan, they are powerless to find any way of defending themselves; and it is a real issue and concern for their families back here in the U.S. So do we have any idea how long this is going to be under review?

Secretary BALLARD. The length of the review, sir, I would like to take for the record and get back to you when that will be completed.

Mr. REYES. Okay.

[The information referred to can be found in the Appendix beginning on page 63.]

Secretary YOUNG. Mr. Congressman, you have raised a very, very difficult and tough challenge for the folks who are forward in the theater. The original contracts that we had in place before the attack prohibited the carrying of any weapons. As we went into this and the contractors, not just the operators of vehicles but other contractors, became at risk, how we try to protect them became the preeminent concern.

And as I mentioned earlier—I think you stepped out—but the concept today is that we integrate those contracted drivers, in the case of convoys, with military escorts as well as military vehicles, so you have a mixed environment. Right now, the current situation, I believe, is as you stated. The contractors in convoys who are operating do not carry weapons. They are integrated with military trucks and military escorts to be able to defend themselves.

The second issue you got to is if something happens and they are being attacked by small arms and are unable to defend themselves is exactly the issue the forward commanders are trying to deal with today. It gets back to the earlier question that was asked on what is the status of a contractor with regard to the Iraqi government with regard to are they part of a military force or are they civilians operating in another country?

So it is a difficult issue, and I think that is driving some of the time that is needed to review this policy. But I can guarantee you that there is no commander forward who is neglecting their duties in trying to protect everybody who is part of the U.S. forces.

Mr. REYES. Oh, sure. And, believe me, that was an assumption that I made, and I have been to Iraq four times, and I have seen nothing to indicate that there is anybody derelict in that. But the simple matter is that they don't know when they are going to get hit, and sometimes when they get hit they get hit with overwhelming force, and they take out trucks, and people are vulnerable. I mean we can't guarantee protection to everyone, but certainly it would be helpful if they had a way—I know if I was driving a truck and got attacked, I would sure feel maybe not safer but at least a little better about having an M–16 or a 45 or something that I could shoot back at somebody trying to level off with a RPG (rocket power grenade) or something like that. So that is important.

The last thing is, Secretary Ballard, when can I anticipate that you will get back to us with the answer that you took for the record?

Secretary BALLARD. As soon as possible, sir.

Mr. REYES. Thank you.

Thank you, Mr. Chairman.

Mr. HEFLEY. Mr. Reyes, I think you raised a very important line of questioning here. The 20/20 Program has a segment called, "Give Me a Break," and I think this would be a good subject for "Give Me a Break." Evidently, our contractors are the only people in Iraq that aren't armed. Everybody is armed in Iraq, and we are taking it under advisement. We have been looking at it for months on end, and I wish you would get back to us with some definitive answer to this and some explanation of why in the world we would have people, civilian or not, there without some means of defending themselves. It just boggles my mind.

Ms. Davis.

Ms. DAVIS OF CALIFORNIA. Thank you, Mr. Chairman. Perhaps I could just follow up on that for a second. Are our military providing force protection, essentially, for contractors? And I think what you are talking about are contractors who are there in support of the military, basically. Is that correct?

Secretary YOUNG. That is correct. My experience over there has been primarily in logistics contract support, so all of the folks that we had contracted were providing direct support to military operations, either driving trucks, providing services on installations, providing other kinds of support for logistics across both Iraq and Kuwait and in Afghanistan.

Ms. DAVIS OF CALIFORNIA. What about other contracts, though, are we also providing force protection for contracts that are let by the CPA?

Secretary YOUNG. I should defer, I think, to Tina for that.

Secretary BALLARD. No, we are not. The contracts require that the contractors provide their own security. In that regard, they provide security for their staff, their workforce, their sites, any of their bed-down areas and their own convoys.

Ms. DAVIS OF CALIFORNIA. And you are saying under no circumstances that you know of are they providing that security, that force protection?

Secretary BALLARD. No. It was very clear in all of our contract solicitations and meetings with industry that the contractors were responsible, in the case of construction, for providing their own security.

Ms. DAVIS OF CALIFORNIA. Thank you. I appreciate that.

Secretary Wynne—and thank you all very much for being here. I wanted to just go back to a few of your written comments. On page 6, you mentioned that the new guidelines are being written, some lessons learned, et cetera; and we were just talking about when the guidelines perhaps for carrying weapons might be available. When do you anticipate the guidelines that you are now writing, as I read this, will be done?

Secretary WYNNE. Well, that is a two-part question. The combatant commanders have authority over the policies relative to carrying weapons in their zone; and we defer to them. So the reason that Secretary Ballard was not clear on when that guidance would be completed is because it is really not in our lane. I mean we are in acquisition and can foster proper contract activity, make sure that people coordinate with the combatant commanders so everybody knows who is who; and that they have done.

Ms. DAVIS OF CALIFORNIA. What about the other issues, though?

Secretary WYNNE. The other issue is more of a coming lessons learned extending, if you will, the Army current practices, asking them if there is a best practice that they have and whether the Navy has a best practice through the contracts that they have laid so that we can consolidate to a single. That could take the better part of this year to conclude.

Ms. DAVIS OF CALIFORNIA. And in the meantime, then, I understand that there are ongoing policies——

Secretary WYNNE. Yes, ma'am, there sure are.

Ms. DAVIS OF CALIFORNIA [continuing]. Which are in place.

Secretary WYNNE. And each of the services has, in fact, policies in place relative to service contracts, whether they are in a combat zone or not in a combat zone.

Ms. DAVIS OF CALIFORNIA. You also, though, talked about the DOD procurement regulations not having a standardized approach——

Secretary WYNNE. Right.

Ms. DAVIS OF CALIFORNIA [continuing]. And, therefore, you are relying on individual contractors to have that. How do you asses then what their standardized approach is?

Secretary WYNNE. We have a whole collection of standardized clauses that we have used over time. And as one of our new clauses becomes widely spread and used, then we will consolidate that to a specific section, I think it is Section M, of the contracting practices where it becomes what the contractors refered to as a boilerplate clause.

In this particular instance, I think each contractor would like to be treated in a fashion that best supports them; and thus far, we have not arrived at a standard clause, so we negotiate from a template, and the template is provided really by the policies and practices that we have.

I would also tell you that, by and large, they are very similar, and one——

Ms. DAVIS OF CALIFORNIA. What would be a red flag?

Secretary WYNNE. One of the examples right now is the insurance premium that we are paying; and it is very prohibitive for small business to get insurance the way we are doing it, because we are buying insurance through the contracts, essentially asking them to go buy insurance.

We found out that there is a best practice going on actually right next door in the USAID agency where they went out and negotiated an insurance premium from competing insurance companies; and now all the contractors buy insurance, if you will, from them. It saves a tremendous amount of money because the pool is larger.

I know that the Army is running a pilot program right now to see whether or not that particular thing could benefit us, and I would say that is an example of one where right now we do not have a standard clause. I think we may get to, with the pilot program's results, to a standard clause.

Ms. DAVIS OF CALIFORNIA. Thank you.

Did you want to comment on that, red flags in terms of these contracts, what it is that you would question, you would ask people to modify their contracts?

Secretary BALLARD. As Secretary Wynne indicated and I indicated in my oral statement, we have found that we do need to have standards. That is why we created the "Contractors Accompanying the Force Provision." In addition to that, in the Army, to create some standard approach to how we work with contractors on the battlefield, we do have a field manual; we have an Army reg that governs how we work with contractors on the battlefield; and, actually, we define the roles of the contractors, describe the relationships between the contractors and the combatant commanders. So we have things in place that the Army follows.

Ms. DAVIS OF CALIFORNIA. Could you comment on salaries? Obviously, these companies are trying to make money on the contracts, and so they have to determine how much they are going to pay their personnel; but what role do you play in that in looking at those, whether it is comparing to domestic, and obviously there are additional costs and certainly liabilities. But how does that work?

Secretary WYNNE. I can tell you that from what I have heard and what I know about supplying personnel, you have to provide a rationale for all of the money that you are going to spend on specific, usually classes of individuals. In service contracts, we attempt to use it by skill level, because we do not want to necessarily tell people what skills they need.

Many times, as a result of the Congressional support, we have gone to performance-based logistics, whereby we will have the contractor actually rely on a certain standard approach. In other words, it will take so many construction engineers, so many designers, so many stevedores, so many truck drivers to do that job. After that point, we have a standard measure against that contract, and we have a risk premium for war zones; we have a risk premium, if you will, for the insurance, and that all stacks up.

Then we follow through with the Defense Contract Audit Agency against some actuals, if you will. When they actually begin to have people flowing into theater, we can then go back and trace just how much is being paid to that individual to make sure that the alleged or the negotiated rates are very similar to the rates being paid.

Sometimes higher, sometimes lower, and that corrective action is put into the next, if you will, rate review that we go on to.

And as far as using performance-based contracts, we contract, for example, to get electricity flowing. If it takes 100 people, that is great; if it takes 1,000 people, that is the way that goes. And we pay sometimes based on the estimate. The situation sometimes changes, as we all know that it has. If it is fixed price, it is fairly painful for the contractor; if it is a cost-plus, then we will go in and we will fully audit all those folks that he has used to make that happen. We make the mission.

And by the way, not just American contractors are involved here. There are many foreign contractors and there are a multitude of Iraqi contractors that are also in this mix.

Ms. DAVIS OF CALIFORNIA. I was going to ask if the new policies would raise the——

Mr. HEFLEY. Ms. Davis, you are out of time here, so——

Ms. DAVIS OF CALIFORNIA. I know. Mr. Chairman, I was just taking your time to continue. I appreciate that, Mr. Chairman. Will the new contracts have any higher threshold for use of Iraqi personnel?

Secretary WYNNE. We actually have encouraged them, because that is the way we want to create Iraqi commerce and business over time. And so we have encouraged the use of Iraqi companies. We have also, by the way, encouraged the use of American small business, which is why we want to make sure that we don't inhibit them from coming by charging them so much.

Ms. DAVIS OF CALIFORNIA. Thank you very much, Mr. Chairman.

Mr. HEFLEY. Mr. Hayes.

Mr. HAYES. Thank you, Mr. Chairman, and I appreciate the opportunity in having this hearing. I particularly appreciate the nature and the spirit of the questions that our friends on the other side of the aisle have asked. I think it is very important.

Secretary Wynne and Young, I enjoyed our time together when we honored Chairman Young the other night; and we are still working on that Navy base for Fort Bragg, Secretary Young.

My question for you is this: Contracting is vitally important in Iraq, and I am a highway contractor in real life, and I have to bid on projects. There is some certainty when I bid on items as to transportation costs, workmen's comp, these kinds of things that determine—would you all spend a little time—I think the public needs to know how uniquely challenging it is, particularly with the need for timing and speed in Iraq, we have a lot of negotiated as opposed to bid contracts. Talk about the challenges of coming up with a contract that you can live with and also the challenges that you have monitoring that.

Secretary WYNNE. I can start. I think Secretary Ballard probably has a little closer experience. As we construct a contract, it is based on a need expressed by the combatant commander in a region; and we try to adhere very closely to that aspect. If you think about it, the Corps of Engineers lets about $6 billion of contracts a year; we are talking about $18 billion in one region of the country, plus there are normal contracting experiences going on. So the folks who express the stress that may be on the contracting activity are to be praised, because there is, in fact, stress on the contracting activity.

There are two sides here. One is you must come up with a contract vehicle that induces competition and induces people to actually want to perform on behalf of the American soldier in this condition and on behalf of our national goals and objectives. That having been said, we are very hard to do business with. We have within our context and contracts very stringent rules, regulations, and follow-up audit rights that are rarely granted in any commercial activity.

So we have induced commercial forums to help us, and many times these construction companies do not often do business with the Department of Defense under a cost-type scheme. When they find out what the follow up is, usually and in the case of even somebody as experienced as KBR, they find their accounting systems to be a little bit wanting as it flows down to subcontractors.

In the case of Iraq, one of the most daunting features is cash payments. They have no banking system, so we don't write checks. As a result, a lot of contracts, especially at the sub-sub level, are cash payments to get the job done.

That having been said, it requires a certain accounting, and our accountants love to—I mean we and our auditors are used to following up with specific rationale for an expenditure of money across the board. And many times in the heat of the zone they are trying to get a job done; you may not have exactly the kind of rationale that pleases a general auditor, and they get to look at it, if you will, under a little bit less trying circumstances than it may have been issued in the first place.

We are actually doing pretty good by rolling up these lessons learned and presenting them in forums to the major contractors that continue to be willing to do business with us; and I think it is getting better and better.

I mentioned before that I reached for a standard. Now, this isn't something I can probably rely on for the future, but when I first asked the question, I was told that in America we question about 3.2 percent of our contract costs; and over in Iraq, we are questioning about 5 percent. Well, against the different standards that apply, I found that to be not an unreasonable risk premium, if you will, but something we had to work on because we would like those standards to be the same, frankly, and so we are working on that.

Mr. HAYES. Anybody else like to comment?

Secretary Ballard.

Secretary BALLARD. Some of the challenges that we have dealt with are the volatility of the environment and the nature of the work that has to be performed. When we did, for example, the solicitations for the construction contract, this was a full and open contract solicitation, and we received 88 bids. We were very pleased to receive 88; but when you consider the breadth of opportunity for offers, 88 is relatively a small number.

We believe that is because the number of contractors who are capable of performing and have the financial solvency to perform and the willingness to go into this environment is very small. And so it is challenging to have contractors who will go into this environment, have the capability to perform in this environment.

Also, the nature of the work is construction, and we all know if we have built anything, that construction has some challenges in it. So there is the staging of the construction projects that also comes into play. That is a challenge. The requirements in the environment are challenging as to how we will do what work, in what order and in what location. So all those things drive the way that we have structured our contracts and the nature of the contracts.

Secretary WYNNE. If I could follow on one thing, and that is since you are in construction, we use British standards. Iraqis are familiar with British standards of construction; and so we enforce British standards, not because we don't have American standards, we do, but they are so familiar with British standards, that when we sent out our quality people, they actually go out with British standards for construction completion. And we do do that.

Mr. HAYES. Thanks for your answers. They have been very, very helpful.

And thank you, Mr. Chairman. I just hope the Iraqis don't have an Internal Revenue Service because cash is necessary, as you say.

Mr. HEFLEY. Mr. Larson.

Mr. LARSON OF CONNECTICUT. Thank you, Chairman Hefley, and thank the panelists for your service to the country and your testimony before the panel here today. I have just a brief statement and then four questions that are a concern of mine as they relate to oversight and the awarding of contracts.

Mr. Wynne, it is my understanding that the U.S. Army Corps of Engineers has primary responsibility for the entire reconstruction effort in Iraq, which, of course, is likely to result in billions of dollars of contracts with hundreds of thousands of Iraqis and foreign nationals under the employ of United States-funded contractors.

Furthermore, it is my understanding that the Army Corps is contemplating sole sourcing insurance contracts under what they call a two-year pilot program. I am very concerned that such an effort would severely distort the insurance marketplace, resulting in the other insurance companies now in the marketplace leaving that marketplace.

Additionally, using the term, "pilot program," I believe misrepresents the huge scope the Army Corps' effort at the risk of liability and having one insurer manage potentially thousands of claims in a very volatile area. I also understand that the pilot program would not only cover contractors in Iraq but would also cover all Army Corps' contracts worldwide.

My colleague in the Senate, the senior senator from my home state, Senator Dodd, along with 4 other senators and over 107 representatives, have requested a study by the GAO on insurance issues in Iraq. I also understand that the CPA Inspector General is conducting its own study of insurance coverage for contract employees in Iraq.

My four questions are as follows: Wouldn't the DOD want the benefit of reviewing the GAO and the CPA's findings in consultation with Department of Labor, the insurance industry, insurance brokers, and other interested parties before proceeding with a formal request for proposal (RFP) for a centrally managed, defense-based insurance program?

Second question, I understand that the DOD intends to move forward and sole source contract all data base administrations (DBA) for DOD employees in Iraq. The question is why would the DOD move forward and award a sole source provider for this insurance before Congress, the Department of Labor, which administers the defense-based insurance program, and DOD have had an opportunity to review the GAO findings?

Third, perhaps I have missed the conclusive study that DOD has conducted to determine that sole sourcing defense-based insurance for all contractors in Iraq is the right thing to do for the U.S. Government, the American taxpayers; and what other alternatives has the DOD explored to address the perceived problems with insurance for its contractors there before going to a sole source contract?

And, finally, does DOD understand the long-term risk to the government in going with a sole source award for a complex insurance line with a long tail exposure? I am told 90 percent of the claims do not develop for at least 7 years. If you could respond.

Secretary WYNNE. First of all, I have a very limited knowledge of the specifics of the pilot program that is being done; however, we get emails like you all say you get emails, and one of the things that we have found anecdotally is sometimes a ten times difference in premium paid between what USAID contractors pay under their contracts and what they have to pay under our contracts. This is extraordinarily daunting to the small business community who wants to come in and play but doesn't have a significant enough pool to leverage the insurance contractors.

So we are leaning toward that way, but I can ask the folks who are doing the pilot program to specifically respond to your question; and we will take that one for the record.

[The information referred to can be found in the Appendix beginning on page 63.]

Mr. LARSON OF CONNECTICUT. Well, I guess my first question is will they—I mean is DOD going to proceed with this before the GAO report and before the CPA's recommendations come back?

Secretary WYNNE. I don't see any reason that we should not proceed with a pilot program. It is not a sole source award, as you said, it is a competitive award done by tens or hundreds of insurance companies, because they all do reinsurance. You were correct in your assessment that a single provider might process the claims, but the claims will be no less daunting than any automotive insurance company in the United States, which, by the way, takes care of a tremendous number of claimants. So I don't see how this distorts the insurance policy to conduct a premium or a small pilot within the Army to find out whether or not—and supply information, by the way, to a GAO study.

That having been said, we are concerned about anything that might distort the market; and so we run a pilot program to determine whether it does or not. I don't think there is—right now, before the GAO study is done, there is no intention to extend this across Army. This is a pilot program that will run for two years.

The folks at Labor, are endorsing the USAID approach to securing insurance, and well they should, if you will. We are looking at that in a cooperative basis, not on a confrontational basis. But that having been said, sir, I will try to make sure that you get a more complete answer from the people who are promoting the pilot program.

Mr. LARSON OF CONNECTICUT. Who is promoting the pilot program?

Secretary WYNNE. I believe it is being done through the U.S. Corps of Engineers.

Mr. LARSON OF CONNECTICUT. So the U.S. Corps of Engineers is promoting this program, even though Congress has gone out and 4 senators and 107 members have signed on to get information before they proceed with the pilot. Is that correct?

Secretary WYNNE. Again, we see no reason to stop a pilot program. That is where we really—from an innovative contracting posture—that is where we really get solid information as to whether to expand it or whether, as you would like to do, restrict it.

Mr. LARSON OF CONNECTICUT. Even if it means risking skewing the whole marketplace?

Secretary WYNNE. Again, I don't believe that the pilot program will have any impact on this huge insurance market that we have. And, besides that, there are, I think, sufficient competitors out there that are willing to take in to perform. I think there is some fear that they will lose their leverage over a small pool of small business people, and perhaps their premiums will go down; but I see that benefiting the American taxpayer.

Mr. LARSON OF CONNECTICUT. Thank you, sir, and I thank the chairman.

Mr. HEFLEY. Mr. Taylor.

Mr. TAYLOR. Thank you, Mr. Chairman, and I thank the gentlemen for sticking around.

A couple quick questions. Have we had any contractors who are either the victims or the cause of friendly fire incidents in the most recent conflict in Iraq and Afghanistan?

Secretary WYNNE. I would have to take that one for the record. I don't know. I did try to find out the number of killed in military action or not killed in military action. I didn't ask for a category like that, I am sorry.

Mr. TAYLOR. General, would you have any knowledge of that?

Secretary YOUNG. No. I, during my time over there, have no knowledge of that, but we need to take that for the record and get that back to you.

Mr. TAYLOR. If a contractor, and this falls into the worst-case scenario, but let's say a contractor gets over there and does something he shouldn't, he robs someone, he rapes a woman, he murders someone other than in the line of duty. To whom is he responsible? Is he under U.S. law? Is he under Iraqi law? Who is the enforcer should this happen? I am not saying that it has happened.

Secretary WYNNE. I think I might ask—perhaps, again, take that one for the record. Let me tell you what I know from my own personal experience, and that is that Iraq is no different than Afghanistan. Afghanistan, by the way, is no different than Japan in all of the husbanding contracts in all of the countries around. We have status of forces agreement, as you well know, sir——

Mr. TAYLOR. Government contractors.

Secretary WYNNE. Government contractors. And the status of forces agreement extends to those government contractors throughout wherever we are. There is a question, which I think is a valid concern that we think we have attributed, which is what is the status of the agreement post-June 30 as far as prior, and we think that the CPA has reached a solid agreement to extend those privileges. I do not know, for example, what is the status of USAID——

Mr. TAYLOR. If I may, let's say you hire someone from Washington, D.C. or Bay St. Louis, Mississippi. As they report to or are on their way to Iraq, are they given instructions telling them under what laws they are going to be bound, what court system they are going to be liable to should something wrong happen?

Secretary WYNNE. I have not seen the orientation program that individual contractors use. The contractor—yes, sir.

Mr. TAYLOR. Okay. Again, they are contractors, at the end of the day, to the United States government. They may be working for Mr. Smith instead of Secretary Wynne, but at the end of the day who's paying the bill is the United States government, the citizens of the United States. What are the rules that we are giving these folks? I would like to know what that is.

Secretary BALLARD. Sir?

Mr. TAYLOR. Yes, ma'am.

Secretary BALLARD. We do have the Military Extraterritorial Jurisdiction Act of 2000, which provides Federal criminal jurisdiction for U.S. contractors performing overseas. This applies to the prime contractors and the subcontractors at any tier.

Mr. TAYLOR. Okay. Are they made aware of this? Is the individual truck driver, is the individual security guard who is hired, is the individual welder who we send over there, is he made aware on the way over there that, "These are the rules we expect you to live by?"

Secretary BALLARD. I would like to take that question for the record.

Mr. TAYLOR. Okay. On one of my trips to the hospital, I encountered an American of Iraqi descent who is a multiple amputee. I don't know the circumstances, but he was badly wounded as a contractor. What requirements, if any, do we as a Nation place on a big company that is hiring this person to provide health care? That person at the moment was being taken care of at one of our military hospitals. Does it become our Nation's obligation; will that person be entitled to Veterans' Administration benefits?

I think that is one of the unknown costs. I am often told that contractors save us money. All I am asking, without making accusations, is are we really comparing apples to apples? We know if a GI becomes a paraplegic or an amputee, we know that we are going to try to fit them with some sort of artificial limb. We know we are going to take care of them the rest of their lives, that is a given.

My question is do our contractors provide the same level of confidence for their employees or do we as a Nation end up footing the bill because the contractor sees them through the minimum amount of assistance and then, in effect, goes into the winds and maybe Medicaid will help them, maybe Medicare will help them, maybe they happen to be fire service and the Veterans' Administration will pick them up. But I have a feeling we end up getting the bill either way as a Nation; and I am wondering if, again, when people are telling me all of the savings by using the contractors, if that is figured in.

And, last, since I do see my red light, in the case of Colombia where a contractor's plane either failed or was shot down, depending on which side of the story you want to believe; but contractors were clearly captured by the FARC (revolutionary armed forces of Columbia) and are still in the FARC's captivity, do we as a Nation require of that contractor, again, the company, not the individual, that that individual's family continues to be paid as long as he is held captive or does the company strike the individual from the roll and the family, Where we, in effect, sent this dad down there to do our Nation's bidding, are they left high and dry?

Secretary WYNNE. To be comprehensive in my response, I think I need to take that for the record. I would only add to the thoughts on insurance that were raised earlier, I am sure that is a concern, because each company now negotiates its own, if you will, policy with insurance companies. I think that is one of the things we would actually like to standardize in our policies to make sure that they are taken care of to the maximum extent possible.

Mr. TAYLOR. But you have not done that yet.

Secretary WYNNE. As far as I know, sir. I will just have to get back to you. I mean it seems the appropriate thing to do, and I am sure that every one of our contractors has taken the appropriate thing to do; but I am not going to swear to it here. I would rather come back with the specifics on the record for you, because I think your question has tremendous merit and very much more cause for concern than I could give it here.

Mr. TAYLOR. Okay. I understand that and I appreciate that; but in the case of the contract employee in Colombia, he has been in the captivity of the FARC since a year ago February, if my memory serves me. Is his family still being compensated or have they taken the attitude that, "Well, he is not flying for us, so we won't pay him?"

Secretary WYNNE. Again, I would have to take that for the record to answer you very specifically on that instance, because I am just frankly not familiar. I do know that every day we get concerned about American citizens no matter where they are in the world.

Mr. TAYLOR. Okay. I would certainly appreciate an answer to both of those questions in the shortest amount of time practical.

[The information referred to can be found in the Appendix beginning on page 64.]

Secretary WYNNE. Yes, sir.

Mr. TAYLOR. Thank you, Mr. Chairman.

Mr. HEFLEY. Mr. Evans.

Mr. EVANS. Thank you, Mr. Chairman. Under the pilot program for contract security guards at Army installations, the Wackenhut Corporation has been providing security for numerous installations in the set to increase its presence under phase three of this program. Department of Energy's Inspector General revealed in January that this same corporation has severely cheated on anti-terrorism drills at the Y–12 Weapons Complex in Oak Ridge, Tennessee. I have also been apprised that the company's security labs have retaliated against whistle blowers at both nuclear weapons plants and power plants.

I ask you, is this the kind of company we should be keeping entrusted with the guarding of sensitive DOD facilities? What is being done to ensure the highest standards?

Secretary WYNNE. Sir, I am going to take that for the record because of the breadth of the question. That having been said, the standards are very clearly stated in any contract that they must measure up to; and we have contracting officer technical representatives to enforce those standards. But I want to be clear about the specific instance since you referenced a specific company and a specific location. I think it would be more appropriate to make sure that you got a comprehensive response.

[The information referred to can be found in the Appendix beginning on page 66.]

Mr. EVANS. Thank you.

Thank you, Mr. Chairman.

Mr. HEFLEY. Thank you very much, and thank the panel for being with us today. We appreciate your presence, and the committee stands adjourned.

Secretary WYNNE. Thank you, sir.

[Whereupon, at 11:28 a.m., the subcommittee was adjourned.]

APPENDIX

JUNE 24, 2004

PREPARED STATEMENTS SUBMITTED FOR THE RECORD

June 24, 2004

FOR OFFICIAL USE ONLY
UNTIL RELEASED BY THE
HOUSE COMMITTEE
ON ARMED SERVICES

STATEMENT OF

THE HONORABLE MICHAEL W. WYNNE
ACTING UNDER SECRETARY OF DEFENSE
(ACQUISITION, TECHNOLOGY, AND LOGISTICS)

BEFORE THE UNITED STATES HOUSE
COMMITTEE ON ARMED SERVICES

June 24

FOR OFFICIAL USE ONLY
UNTIL RELEASED BY THE
HOUSE COMMITTEE
ON ARMED SERVICES

The Honorable Michael W. Wynne
Acting Under Secretary of Defense
(Acquisition, Technology and Logistics)

Good morning Mr. Chairman, Mr. Ranking Member, and Members of the Committee. Thank you for the opportunity to discuss the role of contractors in the Department, particularly with the role of contractors that support the military forces. As I begin, I want to thank the contractors who support us abroad for their courage under fire and support of our national objectives. Numerous contractor personnel have died for their country, and we appreciate and remember their ultimate sacrifice.

Contractors on the Battlefield – A History

Contractors have historically been an integral part of our battlefield team in defending our nation. They have supported the military in every contingency since the Revolutionary War. General George Washington used contracted civilian wagon drivers to haul supplies. Vendors followed Union Troops and sold needed supplies to them during the Civil War. By WW II, civilian workers, hired either individually or through firms, provided support services in all the theaters of war. In the Korean War, contractors provided services ranging from stevedoring, road and rail maintenance to transportation. By Vietnam, contractors were becoming a major part of logistical capabilities within zones of operation providing construction, base operations, water and ground transportation, petroleum supply and maintenance/technical support for high-technology systems. During the Gulf War, there were contractor employees deployed in support of

U. S. Forces providing maintenance for high-tech equipment in addition to water, food, construction and other services.

The use of contractors in a theater of operation releases military units to concentrate on their core mission. Their partnership has allowed us to fill identified shortfalls and give the supported Combatant Commander an additional logistics option.

Current Defense Reliance Upon Contractor Support

Today, we depend upon government contractors to support myriad aspects of the Department's mission. Contractors play a direct role in the research and development of major weapons systems. They maintain both new and fielded weapons systems, and provide critical support in base maintenance and support services. Everything including high-tech, nano-engineering on classified programs, project planning and support, depot maintenance support for aviation platforms, and hauling away trash from our bases and offices can be, and is often, acquired via contract. In fact, in many cases, contractors provide critically needed service support in offices and programs across the Department.

Contractors are also involved with the Department in prosecuting the war on terrorism in every corner of the globe. The role of contractors within the Department is critical and it is certain that the Department, as currently staffed, needs contract support in order to carry out postwar stability and reconstruction work in Iraq. Contracting will have to continue if we are to be successful as a Department and as a country.

Why does the Department contract for support? Using the support provided by contractors extends the capability of the DoD civilian and military workforce and it allows the Department to focus upon its primary mission of defending the nation and safeguarding our freedom. A key impetus for the Department's general reduction of internal resources has been extensive support of competitive sourcing initiatives which strongly encourage government activities to compete their commercial functions with industry. This is done with the goal of establishing a provider of a service that will enhance the quality, economy and productivity of the government. That provider can often be a government contractor.

Often, the Department contracts for support and capabilities where the Department either does not have the expertise or the organic resources. Contracting can enable the Department to access technology and capabilities that would have been unavailable to the Department, would take an inordinate amount of time to develop internally, or would be prohibitively expensive to develop. Accessing commercially available capability makes sense and ensures that we stay ahead of our adversaries.

In today's current environment the Department buys on a performance or outcome basis. We don't necessarily tell the contractor how to build something; we specify what we want as an outcome or result. Often, this can constitute a services contract requirement. For example, we want to be able to fly an aircraft with 99% availability. The contractor provides the necessary maintenance and support personnel to guarantee the aircraft's availability. Historically, we might have bought engines, spare parts, hourly labor and other consumable items. The contract would have largely been a hardware procurement. Further, on a number of information technology requirements, we contract for computer and connectivity service rather than computer hardware and software. We write the contract to require the contractor to keep us supported with the latest in computer technology. How they accomplish this support is up to the contractor. In the same vein, we used to procure rockets and associated hardware to put satellites into space. The details and specifications to accomplish this were mind-boggling. Now, we put the responsibility on the contractor. We buy capabilities which may constitute a services contract. This is consistent with state-of-the-art commercial practices.

Even as we buy smarter, today's acquisition professionals must work harder than ever to manage rising funding requirements, to execute a growing number of contracting actions, and to administer an expanding range and volume of complex acquisitions, including performance-based contracting and services acquisition. The AT&L workforce, as defined and managed by the DoD, shrank by 10%, from 149,439 people in March 1998 (when it was first defined) to 134, 431 people in September 2003. However, from a workload perspective, the investment account funding (*i.e.* RDT&E, Procurement, and Military Construction funds) increased 54%, from $96.5B in FY 1998 to $148.8B in FY 2003. Similarly, total number of contract actions exceeding $100,000 (*i.e.*, those that require the vast majority of work by acquisition personnel) increased 57% from 120,560 in FY 98 to 189,140 in FY 2003. While the AT&L workforce has been shrinking, we continue to place greater demands on our workforce. To help meet this demand efficiently, we contract for project planning and support as specific needs arise. This is just one example of how the Department leverages contractor support to meet mission requirements.

The amount the Department spends on services further illustrates the growing dependence upon contracted support. In 2001 we spent $153B on goods and services and in 2003 we spent $209B. Of that $209B in 2003, we spent $118.B on services including R&D and construction and $91B on supplies or hardware. Of the $118B, $76B is for services excluding R&D and construction. In fact, over the past five years the Department has consistently spent more on services than on weapons, goods, supplies and equipment. Up until the most recent 5 years, the Department consistently spent more on supplies and equipment than on services.

Contracting with industry, prudently approached in the current environment, can be an effective means of leveraging the government's organic capability.

Guidance

Although policy, regulations, and laws exist relating to contractors on the battlefield, the Joint Staff and my organization are drafting overarching policy and procedures to address management of contractor personnel during contingency operations. These polices and procedures will not only provide new policy to address issues encountered during recent operations but will also serve as a "roadmap" to navigate through the existing policy and procedures available today. We also plan to issue new rules on the proper use of non-DoD contract vehicles that are often used to acquire contractor support during contingency operations.

First, we are developing a DoD Directive (and companion DoD Instruction) for the "Management of Contractor Personnel During Contingency Operations" since existing policy does not capture lessons from recent operations nor does DoD currently have a single directive (as recommended by GAO in GAO-03-695 of June 2003) to address the issues of contractor personnel management. Issues being addressed include:

better integration of contractors into planning; maintaining visibility; force protection; protective equipment; weapons; uniforms; medical, and mortuary coverage. Over 400 comments were received and resolved on the first draft of this directive. We plan to coordinate a final draft DoD directive and instruction through the Military Departments, the Combatant Commanders, and the Defense Agencies by the end of July. Barring any significant changes and after reconciling any final comments we hope to have an approved directive in October of this year.

Second, the resolution of policy issues must be consistent with a proposed Defense FAR Supplement rule on "Contractors Accompanying a Deployed Force," which is simultaneously in process. Currently the procurement regulations do not have a standardized approach for addressing most issues that deal with contractors accompanying the forces engaged in contingency, humanitarian, peacekeeping, or combat operations outside the United States. Therefore, individual contracting offices deal with these issues on a contract-by-contract basis, negotiating unique terms and conditions for each contract. The rule is designed to ensure the uniform treatment of contractors that accompany a deployed force, and to enable combatant commanders to rapidly adjust contract requirements in response to changing conditions on the battlefield. There are a number of significant issues that surfaced through the public comment process. For example, there is significant controversy over the policy of whether contractors should carry weapons and under which circumstances the combatant commander can issue directions to contractors. Direction given by combatant commanders can create contractual problems, such as violation of the Anti-Deficiency Act if funds are not available. We have established an ad hoc committee to resolve these issues as we proceed to publication of a final rule, which is planned also for October of this year.

DoD satisfies its wartime contractual needs in a number of ways. This includes the award of standalone contracts, orders issued under existing umbrella type contracts such as multiple agency contracts and General Services Administration Schedules. Recent experience highlights the need to properly use all of these methods and in particular non-DoD contract vehicles. In order to better ensure that non-DoD contract vehicles are the best solution to satisfy DoD needs and to ensure they are properly used, we plan to issue new guidance. Under the new guidance, DoD contracting officers will have to approve use of non-DoD contract vehicles. This will provide more oversight for requirements that currently may flow directly from the customer such as a program manager to a non-DoD entity such as GSA's Federal Technology Service. Also, the contracting officer will have to consider whether the requirement properly fits within the scope of the intended umbrella contract and whether it offers the best way to satisfy DoD requirements, considering schedule, fees, and other factors.

Conclusion

In closing, contractors are an integral part of the Defense team to meet the many missions of the Department. Besides providing critical weapon systems and supplies, the utilization of contractors to provide services releases military units to concentrate on their core missions. To improve on the management of contractors in the battlefield, we are developing overarching policies and procedures to address the issues associated with contractor personnel in the theater of operations. I appreciate the support the House

Armed Services Committee has provided for the Department's acquisition and technology programs and our soldiers, sailors, airmen, and Marines. I look forward to working with you to make our support to the warfighter even better. Thank you.

NOT FOR PUBLICATION UNTIL RELEASED BY THE
HOUSE ARMED SERVICES COMMITTEE
READINESS SUBCOMMITTEE

STATEMENT OF

MR. JOHN J. YOUNG, JR.
ASSISTANT SECRETARY OF THE NAVY
(RESEARCH, DEVELOPMENT & ACQUISITION)

BEFORE THE

SUBCOMMITTEE ON READINESS

OF THE

HOUSE ARMED SERVICES COMMITTEE

ON

THE ROLE OF CONTRACTORS SUPPORTING MILITARY FORCES

JUNE 24, 2004

NOT FOR PUBLICATION UNTIL RELEASED BY THE
HOUSE ARMED SERVICES COMMITTEE
READINESS SUBCOMMITTEE

Mr. Chairman, distinguished members of the Subcommittee, thank you for this opportunity to discuss the Department of the Navy's policies on contractor support to the warfighter, and our practices in overseeing those contractors.

The Navy and Marine Corps Team is unique among the Services in that our mission, being expeditionary in nature, is focused on operating independently, and our operating culture is based on being able to sustain operations organically in an isolated environment, independent of sustained contractor support. We are often called upon to operate in areas where access may be restricted by friends, withheld by neutrals, or denied by adversaries. Naval Forces must be capable of initiating and sustaining combat operations by sea, land, and air without being limited by a lack of logistics or host nation support. The Navy does not carry a large cadre of contractors aboard ship, and the Marines do not typically take contractors ashore to support their landing forces.

Throughout the history of the Naval Services, we have relied to some degree on the support of contractors to effectively and successfully carry out our mission. As a matter of fact, the first Navy supply officer, Mr. Tench Francis, was a civilian contractor hired to conduct purchasing and supply operations on the Navy's behalf. Additionally, since the Navy and Marine Corps Team must always be ready to protect United States interests worldwide on short notice, we maintain fleet units deployed at all times, and have contract support arrangements, commonly called husbanding contracts, in place at ports around the globe, in lieu of establishing an organic Navy presence in every port we might need to visit. These arrangements are put in place in advance, so that they are available when needed to re-supply our ships and to support in-port operations, on an on-call basis.

Historically, the Navy and Marine Corps relied on an organic capability to sustain operations. We designed and built our own ships and submarines in Naval Shipyards. We acquired, stocked, and distributed millions of individual spare parts, necessitating maintenance of an extended logistics supply chain from warehouses, to depots, to transportation points, to the fleet. During the past century, we came to rely on the shipbuilding industry to design and build our ships and submarines for us. In addition, over the past 10-15 years, we have been moving away from being our own logistics support agent to relying on contractors to perform that function.

In the current environment, our industry partners are supporting our forward deployed military forces primarily in the area of specific weapon system support, either under a contracted logistics support arrangement such as Performance Based Logistics or Contractor Logistics Support, or under a field technical services contract. The performance of the work is monitored by the assigned organization – either the Defense Contract Management Agency or the Navy or Marine Corps end-user at the site where the field technical support is performed. In either case, verification of successful performance of the work is a necessary prerequisite for contractor payment.

The only other major industry partner support we use is in the construction arena. The Naval Facilities Engineering Command (NAVFAC), in 2000, awarded a competitive five-year Emergency Construction Capabilities, Worldwide (CONCAP) contract. The

CONCAP contract is a Cost-Plus-Award Fee, Indefinite Quantity/Indefinite Delivery contract which provides the Navy an immediate response for construction and construction related engineering services in support of contingency and emergency situations. In 2003, CONCAP provided support to the Marine Expeditionary Group and the Seabees for work in Kuwait for construction of retrograde facilities and material logistic support valued at approximately $14.6 million. This contract was also used for emergency response for detention facilities at Guantanamo Bay, Cuba, valued at approximately $110 million and typhoon and hurricane response for approximately $62 million over the past two years.

NAVFAC maintains the organic capability and capacity to manage and oversee all its construction work, either through its Resident Officer-in-Charge of Construction (ROICC) organization or through the assignment of Navy Technical Representatives in locations where a ROICC isn't available. Contractor payment is handled through this contract administration structure, again with payments authorized only after confirmation that the work has been satisfactorily completed.

Finally, in response to growing concerns regarding the acquisition of services, and in implementation of Section 801 of the National Defense Authorization Act for Fiscal Year 2002, we developed and implemented a Management Oversight Process for the Acquisition of Services, and I am the approval authority for the largest acquisitions.

Mr. Chairman, in summary, our industry partners provide essential support to the deployed military forces that enables our forces to focus on their core national defense mission. The Navy and Marine Corps acquisition team is continuing to provide our warfighters the support they need, consistent with responsible management and stewardship. Effective acquisition planning, timely contract execution, and responsible management and oversight have provided the Navy and Marine Corps Team the contractor support needed to accomplish the mission. We are doing everything it takes to make sure our Sailors and Marines are provided with the safest, most dependable, and highest performance equipment available within fiscal constraints, together with the logistics and material support necessary to ensure performance wherever and whenever they are needed.

NOT FOR PUBLICATION UNTIL RELEASED BY THE
HOUSE ARMED SERVICES COMMITTEE
READINESS SUBCOMMITTEE

STATEMENT OF

DR. MARVIN R. SAMBUR
ASSISTANT SECRETARY OF THE AIR FORCE
(ACQUISITION)

BEFORE THE

SUBCOMMITTEE ON READINESS

OF THE

HOUSE ARMED SERVICES COMMITTEE

ON

THE ROLE OF CONTRACTORS SUPPORTING MILITARY FORCES

JUNE 24, 2004

NOT FOR PUBLICATION UNTIL RELEASED BY THE
HOUSE ARMED SERVICES COMMITTEE
READINESS SUBCOMMITTEE

Mr. Chairman, Congressman Ortiz and members of the Committee, thank you for this opportunity to participate in today's discussion of contractor oversight within the Department. I proudly represent the acquisition professionals responsible for developing and delivering the warfighting capability that makes the United States Air Force the dominant air force in the world. I'm also happy to testify to the roles of Air Force contractors, their value to the Department, and where we're getting it right. While there is always room for improvement, our experience with contractors and contracting has been largely positive. My hope is that we can achieve mutual agreement of what needs to be fixed. We are committed to work closely with OSD and Congress to make the necessary corrections.

Before I get too far, I'd like to take a moment to describe some of the hard work being done for you by some of the people I represent. Air Force contingency contracting expertise is in high demand across the Department with our officers leading joint contingency contracting operations in the Balkans, Afghanistan, and Iraq. In calendar year 2003, we sent over 400 contingency contracting officers to 58 locations around the world. So far this year, 117 have already been deployed to 24 locations. Contingency contracting officers deploy as part of our Air Expeditionary Force packages. Once in place, they do the hiring of nationals to provide our warfighters with a myriad of local supplies and services. This approach significantly reduces the size, or footprint, of the deployed force as the supplies and services in question, not to mention infrastructure to support them, do not have to come along for the ride. Through April of this year, in CENTAF alone AF, contingency contracting officers have performed over 11,000 contracting actions valued at over 120 million dollars. The Air Force Team has a solid track record in this area, one we can all be proud of.

When I say the Air Force Team, I want to be clear that I'm referring to more than just our roughly 28,000 uniform Airmen and civilians. Indeed the reason we're here today is to discuss policies related to the third component of our team, namely private contractors. In Air Force acquisition, we buy hardware up to and including major weapon systems, along with many types of services (such as base operating support and logistics support just to name a few) from private contractors. We also contract for support and expertise in our acquisition program offices as well. Whether it is on the buying or selling side of the fence, contractors are therefore an integral part of our team.

A key point I'd like to add is that the team-aspect remains intact when contractors employed by the Air Force deploy to a contingency theater in support of our systems. The contractors themselves, Air Force technical personnel, and quality assurance personnel all go as part of a team. In this way, we are able to ensure good contract performance. Also, we take advantage of the Defense Contract Management Agency and the Defense Contract Audit Agency for management oversight and pricing support, both of these organizations having regional offices in Iraq.

The Air Force has also recognized the growing importance of acquiring "services," and in 2002 we established what is now called the Program Executive Office for Combat Mission Support to provide oversight and management of "services" acquisitions of over 100 million dollars. This office conducts annual program reviews of services within their portfolio and serves as the Source Selection Authority for acquisitions within their portfolio. This approach has been recognized and supported by the DoD Inspector General and recently by Congress. In the Senate's version of the FY05 DoD Authorization Act currently under consideration, the committee urges other departments to review our management structure and strongly consider establishing similar structures.

The Air Force has also developed a Management and Oversight of Acquisition of Services Process at the Service level and requires that all programs develop a similar plan for acquisitions within their purview. The process basically requires acquisition status and strategy reviews during critical stages of the acquisition process, not only prior to the award, but also during transition and at least annually thereafter. Additionally, we've established a multi-functional Services Management Council to work services issues that impact all parties associated with an acquisition. First up on the Council's agenda is to address providing training for Quality Assurance personnel assigned to help assure performance on our services contracts.

Air Force Instruction 63-124 on Performance-based Services Acquisitions now applies to all services contracts over the Federal Acquisition Register's Simplified Acquisition Threshold. This AFI emphasizes the need to focus on performance measurement, performance assessment, and performance management after contract award. These areas are addressed and a performance plan developed as part of every applicable services contract acquisition strategy. In order to assist in this effort, we are also developing a computer-based Performance-based Services Acquisition training tool for use by the acquisition team. The training will include a start out with modules on performance measurement and performance management.

Of some note with regards to today's discussions, the Air Force has also established an Integrated Product Team that is examining our Military Interdepartmental Procurement Request, or MIPR as its commonly known, process. Of interest to us is improving the business decisions that are routinely made on contracting out Air Force requirements through other agencies, such as the General Services Administration. The IPT will propose a re-engineered process and provide recommendations for policy or process changes.

Finally, the Air Force is establishing a Services Community of Practice website wherein members of multi-functional teams can communicate and share ideas, challenges, and best practices.

In closing, I'd like to summarize by saying that we in Air Force Acquisition have recognized the critical importance of contractor oversight and in many ways are already at the leading edge in improving how the department does its business in this area. I look forward to a robust discussion of the matter and thank the committee for this hearing and the time allotted for my statement. I'm happy to answer any questions you might have for me.

RECORD VERSION

STATEMENT BY
MS. TINA BALLARD
DEPUTY ASSISTANT SECRETARY OF THE ARMY
(POLICY AND PROCUREMENT)

BEFORE THE

SUBCOMMITTEE ON READINESS
COMMITTEE ON ARMED SERVICES
UNITED STATES HOUSE OF REPRESENTATIVES

SECOND SESSION, 108TH CONGRESS

ON THE ROLES AND MISSIONS OF CONTRACTORS THAT SUPPORT THE DEPARTMENT OF DEFENSE AND THE MILITARY SERVICES

JUNE 24, 2004

NOT FOR PUBLICATION
UNTIL RELEASED
BY THE COMMITTEE
ON ARMED SERVICES

STATEMENT BY
MS. TINA BALLARD
DEPUTY ASSISTANT SECRETARY OF THE ARMY
(POLICY AND PROCUREMENT)

ON THE ROLES AND MISSIONS OF CONTRACTORS THAT SUPPORT THE DEPARTMENT OF DEFENSE AND THE MILITARY SERVICES

INTRODUCTION

Mr. Chairman and distinguished members of the Subcommittee on Readiness, thank you for this opportunity to report to you on the roles and missions of contractors that support the Department of Defense and the military services, specifically the United States Army. It is my privilege to represent the Army leadership and the military and civilian members of the Army's contracting and acquisition workforce, and the Soldiers who rely on us to provide them with world-class weapon systems, equipment, and services so they can successfully accomplish any mission at anytime, anywhere in the world.

This is a time of tremendous change, and we are most grateful for your wisdom, guidance, and strong support. Over the last year, our Army has met the demands of the Global War on Terrorism with more than 330,000 troops deployed around the world in more than 120 countries. Our Army was instrumental in the defeat of Saddam Hussein and the subsequent liberation of more than 46 million people from oppression and despair. Our Army remains a central and critical participant in Operation Iraqi Freedom (OIF) and Operation Enduring Freedom (OEF) in Afghanistan. Our contracting workforce is at the forefront in the reconstruction and humanitarian relief effort in Iraq and I would like to briefly highlight the Army's significant role.

contracts. A comprehensive "Contractors Accompanying the Force Guidebook" was developed as a companion document to provide operational details, and it included sample language that could be tailored for individual needs. It's clear that contractors who accompany the Army in theater need to be completely aware of the full spectrum environments in which they will likely operate. The proper mechanism for this information is the contract; and the purpose of the clause and "Guidebook" is to provide contractor deployment information in Army contracts accurately and consistently. It is also designed to enlighten the operational community, including field commanders and logisticians, on what they can expect from support contractors. In addition, the Army has published a field manual (FM 3-100.21) entitled "Contractors on the Battlefield." The Army takes great pride in a report dated June 24, 2003, from the General Accounting Office stating that of the four military services, only the Army has developed substantial guidance for dealing with contractors.

In the Army's continued efforts to improve management of service acquisitions, the Army Services Strategy Panel (ASSP), which I chair, is focused on the importance of developing and maintaining sound acquisition strategies to ensure services are properly planned, based upon clear, performance-based requirements with required outcomes that are identified and measurable, and acquired by smart business arrangements. ASSPs will be conducted as early as possible in the acquisition planning process to develop a systematic and disciplined approach to achieve an affordable, efficient, and effective acquisition.

Another issue of key importance in maintaining oversight and management of contractors is the Contracting Officer's Representative (COR). A COR is an authorized representative of a Contracting Officer who assists in the technical monitoring or administration of a contract. In the Army, the COR must be a Government employee who has been trained as a COR and authorized by the Contracting Officer to act in that capacity. The COR has several major duties and responsibilities, including the following:

- verify that the contractor performs the technical requirements of the contract in accordance with the contract terms, conditions, and specifications;
- emphasize the quality provisions for both adherence to the contract provisions and to the contractor's own quality control program;
- perform inspections and verify that the contractor has corrected all deficiencies;
- maintain liaison with the Contracting Officer and direct communications between the Contracting Officer and the contractor.

CONCLUSION

I have a deep appreciation for the courage, competence, dedication, and hard work of our contracting personnel. Their work has been – and continues to be – performed in accordance with proper procedures, by military service members and civilians who are concerned and committed to accomplishing their work under the toughest and most austere conditions.

Our contractors and contracting personnel are often the unsung heroes of any deployment. Their customers are the warfighters – the men and women who depend on them to do their jobs. The success of our warfighters is linked directly to the success of the contracting workforce. We must continue to work hard and work together with industry to develop win-win situations where industry makes a fair profit and our warfighters get better products and services to accomplish their missions quickly and decisively and return home safely.

QUESTIONS AND ANSWERS SUBMITTED FOR THE RECORD

JUNE 24, 2004

QUESTIONS SUBMITTED BY MR. HEFLEY

Mr. HEFLEY. What is the current policy with respect to determining what functions should or should not be performed by a contractor?

Secretary WYNNE. The Department's manpower policy provides criteria for determining whether, and under what circumstances, functions must be performed by government (military or civilian) personnel. Functions that the Department has identified as "inherently governmental" are not suitable for performance by contract. In addition, the Department may exempt from private sector performance, based on law, regulation, or international agreement, or to provide for overseas rotation, career progression, wartime assignments, continuity of infrastructure operations, and other similar requirements. When making these decisions, Defense officials must often consider not only the nature of the work, but also the circumstances and risks. Additionally, as threat levels and risks change, decisions as to what should and should not be contracted may change. Defense policy requires that risk mitigation take precedence over cost savings when there are risks to national defense, readiness, or government responsibilities. The Department uses the manpower criteria in determining the mix of military and civilian manpower and contract support for the operating forces and infrastructure, and when considering options for converting from one source of performance to another during the development of the annual personnel authorization request to Congress, as required by section 129a of title 10, United States Code.

Mr. YOUNG. Functions that are inherently governmental should not be performed by a contractor. Inherently governmental functions are those functions that are so intimately related to the public interest as to mandate performance by Government employees, and include activities that require either the exercise of discretion, in applying Government authority, or the making of value judgments in making decisions for the Government. Functions that are not inherently governmental should be evaluated on a case-by-case basis to determine the benefits and risks of contractor performance. Considerations vary from one situation to the next, and include where the function is to be performed, whether the requirement for the function is permanent or short-term, etc.

Dr. SAMBUR. The Air Force must concentrate its resource investments in areas most directly related to accomplishing its core missions to sustain readiness, improve performance by doing business more efficiently and cost-effectively, and to generate funds for force modernization. To do so, the Air Force will not contract inherently governmental, military-essential or legislatively exempt activities. Activities that are commercial are competitively sourced using thye procedures in OMB Circular A–76 ("Revised") to determine whether the activity should be contracted, performed by government personnel, or by a Service Agreement.

Ms. BALLARD. Contractor personnel may perform a great many of the tasks normally associated with the duties of both Military and Civil-Service personnel. Or, a company may be tasked to provide a specified service due to the non-availability of an appropriate unit in the supported organization. However, a contractor cannot be required to perform an "inherently governmental function" such as supervising Government Employees or Military Service members, collecting or distributing funds or currency, or conducting combat operations.

Examples of requests for contracted support which were disapproved on the basis of inherently governmental duties include: a request for a contractor to collect old currency and transport and distribute new currency during operations in Afghanistan; a request for a contractor to handle the sale of Iraqi oil and to collect the revenue from these sales and transfer the money to the Iraqi government; and, several requests to arm contractor personnel with offensive weapons in order to provide their own Force protection, something that is a specified mission of the supported commander.

Mr. HEFLEY. In a June 2003 report, "Military Operations: Contractors Provide Vital Services to Deployed Forces but are not Adequately Addressed in DOD Plans," the Comptroller General recommended that DOD develop guidance and doctrine on how to manage contractors that support deployed forces and standardize deployment language for contracts that support or may support deployed forces.

a. Has this been done?
b. If so, explain further actions taken.
c. If not, please explain why.

Secretary WYNNE. The Department of Defense (DoD) taken several steps to address this recommendation. DoD published a proposed rule in the Federal Register for a clause in the Defense Federal Acquisition Regulation Supplement that would standardize language used in contracts to support deployed forces. The Department is currently reviewing the public comments received in response to the proposed rule. In addition, the DoD is developing a draft directive and a draft instruction that will set out DoD's policies regarding contractors supporting deployed forces.

Mr. HEFLEY. The Air Force has a Program Executive Office for Combat Mission Support to provide oversight and management of service acquisitions over $100 million. What is the benefit of this organization and have the other services considered such an office?

Secretary YOUNG. We defer to the Air Force to describe the be fits of their Program Executive Office for Combat Mission Support. Within the Department of the Navy, in the implementation of Section 801 of the FY 2002 National Defense Authorization Act, we concluded that the best approach for us was one which held program managers and program executive officers accountable for all their acquisitions, services and otherwise. This was accomplished within the already existing structure established for management of systems acquisition, but tailored to reflect the importance of improved attention to significant services efforts. Thus, we developed and implemented a Management and Oversight Process for Acquisition of Services in which strategies for acquiring services need to be reviewed and approved at progressively higher levels of leadership as the dollar value of the services effort increases.

Mr. HEFLEY. The Air Force has a Program Executive Office for Combat Mission Support to provide oversight and management of service acquisitions over $100 million.

What is the benefit of this organization and have the other services considered such an office?

Dr. SAMBUR. The Air Force Program Executive Officer for Combat Mission Support provides a number of benefits. The primary benefit is that it provides a single focal point for all service contracts, as intended by FY02 NDAA, Section 801. The Air Force Program Executive Officer for Combat Mission Support serves as the source selection authority for services valued above $100 million and A–76 Studies with more than 300 full time equivalents. This office monitors the execution of service contracts using various performance measures.

The other services have considered such an office. The Air Force Program Executive Officer for Combat Mission Support has met with the Army to discuss this organization and lessons learned since its inception in March 2002.

Mr. HEFLEY. What are the lessons learned from Iraq and Afghanistan about the use of . contractors on the battlefield?

Ms. BALLARD. Collecting lessons learned from Iraq and Afghanistan about the use of contractors on the battlefield is a continuous process. The following are examples of some of the more valuable lessons that we've learned so far:

Having contractors embedded with units greatly reduced down time and eliminated the challenge of transporting contractors from Point A to Point B to catch up to the unit experiencing a problem. This also improved force protection by reducing the amount of time contractors were put at risk traveling in convoys.

Integrating the contractors with the unit early in the training and equipping cycle, and deploying them with the unit helped develop habitual relationships and improved coordination and understanding. Cross-training support contractors provided faster troubleshooting analysis and reduced the logistics footprint as desired by senior Army leadership.

With the increased use of contractors in contingencies, the Army recognized the need to account for its contractors and their movements. Thus, the Army Materiel Command Logistics Support Element-Southwest Asia took the lead to create a contractor coordination cell (3–C) to help account for Army contractors in support of the Coalition Forces Land Component Command (CFLCC) and their movements during Operation Iraqi Freedom (OIF).

Other lessons learned include the need for consistent guidance for contractors deploying to Iraq; taking time to hold a partnering session at the onset of any project greatly improves effective communication between the Government and the Contractor, which is particularly important on a fast-track design/build project where many decisions have to be made and agreed upon which affect the timely procurement and delivery of materials and equipment, and thus the overall schedule; the schedule for individual task orders, including completion dates needs to be analyzed in terms of reasonableness and practicality, as well as consideration of security issues; having a contractor liaison/senior representative co-located with the Task Force Headquarters greatly improves communication, information flow and responsiveness, while freeing up those in the field offices to concentrate on day to day operations and construction management; wrapping the contract definitization process into a short training session at the start of the contracting process would enable all parties, Government and contractor, to understand the unique process; the Government and contractors should have routine program reviews with each other, both in the field and stateside in order to identify issues and problems and keep everyone informed at the appropriate level; and, it is important to have DCAA involved from the start of the contracting process.

Mr. HEFLEY. The Air Force has a Program Executive Office for Combat Mission Support to provide oversight and management of service acquisitions over $100 million. What is the benefit of this organization and have other offices considered such an office?

Ms. BALLARD. Last year, the Deputy Under Secretary of the Army asked if the Army should pursue a Program Executive Office (PEO) like concept for acquisition of services, similar to the organization established by the Air Force. We briefed the Army Acquisition Executive (AAE) in May 2003 and recommended that Army not pursue consideration of a PEO concept as the Army process and organization for managing services was in its infancy stage and needed time to mature. We suggested and the AAE agreed with our recommendation that we revisit the Army approach after we had time to review and collect appropriate data on our process.

The Army is strategically managing and overseeing services both organizationally and by process. From an organizational standpoint, the Army Contracting Agency was specifically created to provide opportunities to leverage buying power, and reduce cost for acquisition of general information technology, professional and installation support services. Process wise, the Army is overseeing the acquisition of services through integrated reviews of proposed acquisitions by me and other senior Army leaders via Acquisition Service Strategy Panels (ASSPs). The Panel reviews provide the forum to ensure that service acquisitions are acquired through a strategic approach and consist of business arrangements that are in the best interest of the Army and the Department of Defense.

The review panels have been ongoing for approximately one year and at least eight Headquarters Department of Army level panels have been conducted for requirements with an estimated value in excess of $500M.

QUESTIONS SUBMITTED BY MR. JONES

Mr. JONES. I understand that over 107 Representatives and 5 Senators have requested a study by the GAO on insurance issues in Iraq. Wouldn't the DoD want the benefit of reviewing the GAO's findings in consultation with the Department of Labor and the insurance industry before proceeding with the RFP? Shouldn't the GAO's findings be vetted with the insurance industry to ensure a comprehensive review of the facts?

Secretary WYNNE. We have encouraged the U.S. Army Corps of Engineers (USACE) to pursue a Defense Base Act (DBA) workers' compensation insurance pilot program because we received complaints last year from companies doing business in Iraq (and from the Professional Services Association) that: (1) the rates for this mandatory insurance had increased precipitously—going from $4 to over $20 per $100 of labor costs, and (2) in some cases, they could not get DBA insurance at all. This difficult DBA insurance market hit small businesses particularly hard because there is often a minimum premium of $15,000, regardless of how few contractor employees are overseas or how short a time period they will be there.

The USACE pilot program we envision is based on two similar, successful programs at the Agency for International Development (AID) and the Department of State (DOS). AID has been awarding its competitive requirements contracts for DBA insurance since the late 1970s, while the DOS program began in 1992—with regular recompetitions for these contracts every five years. The DBA insurance rates under both the AID and DOS programs have been very reasonable, with no minimum premium requirements.

It is incorrect to categorize the planned USACE pilot program as a monopoly or sole source procurement. As with the DBA programs at AID and DOS, the USACE pilot program contract will be competed. After award, all requirements for DBA insurance on USACE contracts will be placed with the winning contractor for an initial period of two years. This is no different from what DoD does on other competitive requirements-type contracts. If the USACE pilot program proves successful, it will be periodically recompeted. We also want to emphasize that this is not a DoD-wide program; it will be limited to the DBA insurance requirements of USACE contractors only.

USAGE currently plans to issue a draft Request For Proposals (RFP) for its DBA insurance pilot program and then to carefully consider any industry comments before issuing a final RFP. If the proposals submitted for the pilot program are not advantageous to the Government based on a comparison with open market DBA rates and availability, USAGE will not award a contract. However, in view of the longstanding, successful programs at AID and DOS, we expect that the USAGE pilot program will also prove successful.

We see no reason to delay the USAGE pilot program for the ongoing GAO review. The goal of this initiative is to determine whether or not it would be advantageous for USAGE to competitively award a no-cost requirements contract to an insurance company to provide DBA insurance for all its contractors with overseas employees, as has been done successfully for many years at AID and DOS. This determination will be based on a careful comparison of the proposals submitted in response to the pilot program RFP with open market DBA rates and availability to obtain the best value for the taxpayers.

Mr. JONES. Given the numerous concerns raised by the insurance industry and the insurance brokerage industry over a centrally managed program, does the DoD plan to conduct an industry forum/round table before proceeding with a formal RFP?

Secretary WYNNE. USACE plans to conduct an industry forum after issuing a draft RFP, in order to obtain input from the insurance industry prior to finalizing it.

Mr. JONES. What is DoD's timeline for issuing the RFP?

Secretary WYNNE. USACE's current timeline for its DBA insurance pilot program is:

July 30: Release draft RFP
August 13: Industry Forum
August 27: Issue Best Value RFP
October 30: Contract Award

Mr. JONES. Is the DoD now working with the insurance industry to explore alternatives to a centrally managed program?

Secretary WYNNE. Based on the longstanding success of the competitive DBA insurance programs at AID and DOS, this approach appears to be the best option for USACE to explore at this time.

Mr. JONES. In light of the anecdotal evidence that some small business defense contractors are having difficulty in obtaining DBA insurance, is the DoD considering targeted solutions to specifically address this narrow problem?

Secretary WYNNE. DoD's concerns regarding current DBA rates and availability are broader than just the problems encountered by small businesses in obtaining coverage, and we believe those concerns may best be addressed by a competitive DBA insurance program similar to the very successful ones at AID and DOS.

Mr. JONES. Given the Department of Labor's administrative responsibility for DBA insurance, is the DoD working closely with the Department of Labor to analyze the DBA insurance market? Does DoD also plan to coordinate any new program initiatives with the Department of Labor?

Secretary WYNNE. Historically, the Department of Labor (DOL) has encountered no significant problems in carrying out its administrative responsibilities under the DBA on the AID and DOS programs. We have discussed the USACE pilot program with DOL, and it has expressed no objections to this initiative. Of course, any contract under the USACE pilot program will only be awarded to an insurance company that has been duly authorized by DOL to provide this type of mandatory insurance.

QUESTIONS SUBMITTED BY MR. CALVERT

Mr. CALVERT. Will the implementation of the GAO recommendations for improving how the Department of Defense addresses contractor support in contingency plans be crafted to comport with the Department's transformational initiatives? Is there enough flexibility in how you will address the noted deficiencies that you won't be forced to change course for every contingency that we face in the future?

Secretary WYNNE. The Department of Defense (DoD) has taken several steps to address the GAO recommendations concerning contractor support in contingency operations. DoD published a proposed rule in the Federal Register for a clause in the Defense Federal Acquisition Regulation Supplement that would standardize language used in contracts to support deployed forces. The Department is currently reviewing the public comments received in response to the publication of this proposed clause. In addition, the Department is developing a draft directive and a draft instruction that will set out the Department's policies regarding contractors supporting deployed forces. When finalized, these documents will offer sufficient flexibility to ensure smooth implementation of contract requirements in any future contingency operation.

QUESTIONS SUBMITTED BY MR. REYES

Mr. REYES. One issue that I have not received a response from, and this is predicated from concerns that have been expressed to me by sometimes parents and sometimes spouses of contractors that are actually working in Iraq for Brown and Root or Haliburton largely in those convoys. Are they allowed now to carry weapons, because based on the questions that I have been asked, the general consensus is that contractor employees in those situations are prohibited from being armed because of, I guess, violation of Iraqi law?

Well, you know, that policy has been under review now—because we had General Meyer here well over a month ago, I guess—and that was the response then. How long do you anticipate the review process to take place?

Ms. BALLARD. The Office of the Chairman of the Joint Chiefs of Staff provided a response to this question to the Office of the Secretary of Defense. Their answer will be provided with Mr. Wynne's responses to Questions for the Record.

QUESTIONS SUBMITTED BY MR. LARSON

Mr. LARSON OF CONNECTICUT. My four questions are as follows: Wouldn't the DOD want the benefit of reviewing the GAO and the CPA's findings in consultation with Department of Labor, the insurance industry, insurance brokers, and other interested parties before proceeding with a formal request for proposal (RFP) for a centrally managed, defense-based insurance program?

Second question, I understand that the DOD intends to move forward and sole source contract data base administrations for DOD employees in Iraq. The question is why would the DOD move forward and award a sole source provider for this insurance before Congress, the Department of Labor, which administers the defense-based insurance program, and DOD have had an opportunity to review the GAO findings?

Third, perhaps I have missed the conclusive study that DOD has conducted to determine that sole sourcing defense-based insurance for all contractors in Iraq is the right thing to do for the U.S. Government, the American taxpayers; and what other alternatives has the DOD explored to address the perceived problems with insurance for its contractors there before going to a sole source contract?

And, finally, does DOD understand the long-term risk to the government in going with a sole source award for a complex insurance line with a long tail exposure? I am told 90 percent of the claims do not develop for at least 7 years. If you could respond.

Secretary WYNNE. We have encouraged the U.S. Army Corps of Engineers (USACE) to pursue a Defense Base Act (DBA) workers' compensation insurance pilot program because we received complaints last year from companies doing business in Iraq (and from the Professional Services Association) that: (1) the rates for this mandatory insurance had increased precipitously - going from $4 to over $20 per $100 of labor costs, and (2) in some cases, they could not get DBA insurance at all. This difficult DBA insurance market hit small businesses particularly hard because there is often a minimum premium of $15,000, regardless of how few contractor employees are overseas or how short a time period they will be there.

The USACE pilot program we envision is based on two similar, successful programs at the Agency for International Development (AID) and the Department of State (DOS). AID has been awarding its competitive requirements contracts for DBA insurance since the late 1970s, while the DOS program began in 1992—with regular recompetitions for these contracts every five years. The DBA insurance rates under both the AID and DOS programs have been very reasonable, with no minimum premium requirements.

It is incorrect to categorize the planned USACE pilot program as a monopoly or sole source procurement. As with the DBA programs at AID and DOS, the USACE pilot program contract will be competed. After award, all requirements for DBA insurance on USACE contracts will be placed with the winning contractor for an initial period of two years. This is no different from what DoD does on other competitive requirements-type contracts. If the USACE pilot program proves successful, it will be periodically recompeted. We also want to emphasize that this is not a DoD-wide program; it will be limited to the DBA insurance requirements of USACE contractors only.

USACE currently plans to issue a draft Request For Proposals (RFP) for its DBA insurance pilot program and then to carefully consider any industry comments before issuing a final RFP. If the proposals submitted for the pilot program are not advantageous to the Government based on a comparison with open market DBA rates and availability, USACE will not award a contract. However, in view of the longstanding, successful programs at AID and DOS, we expect that the USACE pilot program will also prove successful.

We see no reason to delay the USAGE pilot program for the ongoing GAO review. The goal of this initiative is to determine whether or not it would be advantageous for USAGE to competitively award a no-cost requirements contract to an insurance company to provide DBA insurance for all its contractors with overseas employees, as has been done successfully for many years at AID and DOS. This determination will be based on a careful comparison of the proposals submitted in response to the pilot program RFP with open market DBA rates and availability to obtain the best value for the taxpayers.

The USAGE pilot program will not result in a sole source contract. If it is determined to be advantageous to the Government, USAGE will competitively award a requirements contract similar to those awarded for the longstanding, successful programs at AID and DOS. Historically, the Department of Labor (DOL) has encountered no significant problems in carrying out its administrative responsibilities under the DBA on the AID and DOS programs. We have discussed the USAGE pilot program with DOL, and it has expressed no objections to this initiative. Of course, any contract under the USAGE pilot program will only be awarded to an insurance company that has been duly authorized by DOL to provide this type of mandatory insurance.

The USAGE pilot program is the DoD study of this issue. Based on the longstanding success of the competitive DBA insurance programs at AID and DOS, this approach appears to be the best option for USAGE to explore at this time.

Under the DBA, the contractor and its DOL-authorized insurance provider bear the risk for workers' compensation claims, not the Government.

QUESTIONS SUBMITTED BY MR. TAYLOR

Mr. TAYLOR. Have we had any contractors who are either the victims or the cause of friendly fire incidents in the most recent conflict in Iraq and Afghanistan?

Secretary WYNNE. To date, one contractor personnel has been killed by friendly fire.

Mr. TAYLOR. If a contractor, and this falls into the worst-case scenario, but let's say a contractor gets over there [to Iraq] and does something he shouldn't, he robs someone, he rapes a woman, he murders someone other than in the line of duty. To whom is he responsible? Is he under U. S. law? Is he under Iraqi law? Who is the enforcer should this happen?

Secretary WYNNE. Department of Defense (DoD) employees, DoD contractors (including subcontractors at any tier), and their employees, are subject to U.S. law for any felony-level offense committed while employed by or accompanying the armed forces outside the United States. Congress has made several, but not all, U.S. Federal criminal laws extraterritorial and, as such, that they apply beyond the shores of the United States.

In 2000, the Military Extraterritorial Jurisdiction Act (MEJA, sections 3261-3267 of tile 18, United States Code) was enacted to ensure that all felony offenses committed by DoD employees and DoD contractors while outside the United States are now subject to criminal prosecution in the U. S. District Courts.

Without protections afforded by international legal agreements or other arrangements with the host country, contractors deployed overseas are subject to the criminal laws of the country in which they are deployed. Coalition Provisional Authority (CPA) Order 17 (Revised) is such an agreement promulgated after consultations with the incoming Iraqi Interim Government.

CPA Order 17 (Revised) defines "contractors" (including employees and subcontractors) as non-Iraqi individuals not normally resident in Iraq who supply goods and services under a contract or grant agreement to or on behalf of the Multinational Force (MNF).

Only contractors who are DoD employees or DoD contractors are subject to MEJA's criminal jurisdiction. All other contractors in Iraq are not subject to MEJA's criminal jurisdiction, but they may be subject to their sending state's criminal laws and Iraqi criminal jurisdiction.

Under CPA Order 17 (Revised), MNF contractors are immune from arrest, detention or legal proceedings in Iraqi courts or other Iraqi legal bodies, but only with respect to acts performed by them pursuant to the terms and conditions of the contract or subcontract. (Emphasis added.) Criminal conduct of the nature referenced would not constitute such acts and, as such, contractors would not be immune from Iraqi criminal jurisdiction.

Absent the approval of the Attorney General or Deputy Attorney General of the United States, there can be no prosecution under MEJA, if a foreign government, in accordance with jurisdiction recognized by the United States, has prosecuted or is prosecuting such person for the conduct constituting such offense.

U.S. Attorney Offices enforce the law and prosecute MEJA cases. The Department of Defense and other Department of Justice entities (e.g., FBI; U.S. Marshals Service) provide investigative, logistical, and other support to U. S. Attorneys, as necessary. MEJA jurisdiction currently exists and has twice been used in the United States to indict offenders.

Proposed DoD regulations to implement MEJA procedures at overseas locations have been drafted and are currently being reviewed by the Senate and House Judiciary Committees.

Mr. TAYLOR. Is the individual truck driver, is the individual security guard who is hired, is the individual welder who we send over there, if he made aware on the way over there that, "These are the rules we expect you to live by?"

Secretary WYNNE. While our Judge Advocate General (JAG) representatives brief that "possible criminal action may result" from violations of the Laws of War, the Military Extraterritorial Jurisdiction Act (MEJA) 2000 act is not mentioned specifically in the block of instruction for individual contractors deploying to the Middle East. We brief Law of Land Warfare and the Joint Chiefs of Staff (JCOS) Rules of Engagement in accordance with TRADOC standards and guidelines. We also cover the 1996 War Crimes Act and how it applies to civilian contractors.

Mr. TAYLOR. In the case of Colombia where a contractor's plane either failed or was shot down, depending on which side of the story you want to believe; but contractors were clearly captured by the FARC (revolutionary armed forces of Columbia) and are still in the FARC's captivity, do we as a Nation require of that contractor, again, the company, not the individual, that that individual's family continues to be paid as long as he is held captive or does the company strike the individual from the roll and the family, where we, in effect, sent this dad down there to do our Nation's bidding, are they left high and dry?

Secretary WYNNE. All Department of Defense (DoD) contractors who perform contracts outside of the United States must provide the workers' compensation insurance mandated by the Defense Base Act (DBA) for their overseas workers. Contractors may satisfy this statutory requirement either by purchasing DBA insurance from a firm approved by the U.S. Department of Labor (DOL) or by applying to DOL for approval to self-insure. Failure to secure the required DBA insurance may result in the contractor's CEO and other corporate officials being found personally liable for satisfying any worker compensation claims.

Every overseas contractor employee covered by the DBA is also deemed covered by the War Hazards Compensation Act (WHCA). The WHCA protects these employees for war hazards such as capture, injury, or death and was intended by Congress to cover those situations that are not compensable under the provisions of the DBA. WHCA claims are paid or reimbursed by DOL from the Employees Compensation Fund established by the Federal Employees Compensation Act; therefore, there is no requirement for contractors to secure insurance coverage for such risks.

Under the DBA and the WHCA umbrella of workers' compensation protection established by Congress, overseas contractor employees who are either injured on the job or fall victim to a war hazard are entitled to health and disability benefits at a marketplace level of care, in a medical facility of their choosing. They would not be entitled to any Veterans' Administration benefits solely as a result of their being injured employees of a DoD contractor. The amputee of whom Congressman Taylor spoke should be entitled to either DBA or WHCA benefits, depending on the particular circumstances of his injuries.

Mr. TAYLOR. In the case of the contract employee in Colombia, he has been in the captivity of the FARC since a year ago February, if my memory serves me. Is his family still being compensated or have they taken the attitude that, "Well he is not flying for us, so we won't pay him?"

Secretary WYNNE. Contractor employees who were killed or captured by the Revolutionary Armed Forces of Columbia (FARC) in Colombia involves a Department of State, rather than a DoD, contract, but we understand that War Hazards Compensation Act (WHCA) death and detention benefits claims have been filed for them with the Department of Labor.

QUESTIONS SUBMITTED BY MR. EVANS

Mr. EVANS. Under the pilot program for contract security guards at Army installations, the Wackenhut Corporation has been providing security for numerous installations in the set to increase its presence under phase three of this program. Department of Energy's Inspector General revealed in January that this same corporation has severely cheated on anti-terrorism drills at the Y–12 Weapons Complex in Oak Ridge, Tennessee. I have also been apprised that the company's security labs have retaliated against whistle blowers at both nuclear weapons plants and power plants.

I ask you, is this the kind of company we should be keeping entrusted with the guarding of sensitive DOD facilities? What is being done to ensure the highest standards?

Secretary WYNNE. We have reviewed the Department of Energy (DOE) Inspector General (IG) Report Number DOE/IG-0636, "Protective Force Performance Test Improvements," dated January 2004. The Army has not had such performance problems with Wackenhut under its guard services contracts.

The Army's Northern Region Contracting Center contract is for gate guards, and perimeter security patrols at individual posts throughout the United States. It was awarded to Alutiiq Corporation with Wackenhut as a subcontractor. Alutiiq is responsible to the Army for performance. All work by Wachenhut is monitored by Alutiiq and government employees. Wackenhut was evaluated prior to subcontract award and was found to have extensive successful past performance in providing these types of security services. Wackenhut also was determined to be financially sound.

The contract is a firm fixed price, award fee contract. The award fee provides an incentive for good performance. Performance is evaluated against a detailed Quality Assurance Surveillance Plan (QASP). Performance factors such as positive image to the U.S. Army, fully trained and staffed personnel, and maintenance of control of installation entry points are reviewed on a daily basis. The findings from this QASP evaluation are provided to an award fee board, comprised of senior Army officials in Washington D.C., and are used as the basis for determining the amount of fee to award the contractor. An award fee board and the Award Fee Determining Official (AFDO) have evaluated contractor performance at Fort Bragg and Carlisle Barracks and performance to date has been above the satisfactory rate sufficient to justify an award fee of 100 percent at Fort Bragg and 97 percent at Carlisle Barracks.

Ms. BALLARD. The contract in question was written for gate guards, and perimeter security patrols at individual posts throughout the United States. A contract was awarded to Alutiiq Corporation of which Wackhenhut is a subcontractor. There is no contract with Wackenhut Corporation. Alutiiq, not Wakenhut, is responsible for performance. Wakenhut employees do not perform duties independently. They work side by side with Alutiiq employees. As a result, their work is constantly monitored by Alutiiq and government employees. Any real or perceived performance risk is mitigated through this arrangement. Wackenhut was also evaluated prior to award as a subcontractor and found to have extensive successful past performance in providing these types of security services. Wackenhut also was determined to be financially sound. Wackenhut is not on the listing of contractors debarred from doing business with the government.

The guard services contract is a firm fixed price, award fee contract. The award fee provides an incentive for maximum performance. Performance is monitored by a detailed Quality Assurance Surveillance Plan (QASP). Performance indicators such as positive image to the U.S. Army, fully trained and staffed personnel, and maintaining control of installation entry points are reviewed on a daily basis. The findings from this QASP are provided to an award fee board, comprised of senior level Army officials in Washington D.C., and are used as the basis in determining the amount of fee to award the contractor. An award fee board has already convened to evaluate contractor performance at Fort Bragg and Carlisle Barracks and provided their findings to the Award Fee Determining Official (AFRO). The board and AFDO found performance to date has been above the satisfactory rate; awarding an award fee of 100% at Fort Bragg and 97% at Carlisle Barracks.

○